The Almagest:
Introduction to the Mathematics of the Heavens

The Almagest:
Introduction to the Mathematics of the Heavens

Claudius Ptolemy

Selections translated by Bruce M. Perry
Edited with notes by William H. Donahue

Green Lion Press
Santa Fe, New Mexico

Manufactured in the United States of America

Published by Green Lion Press, Santa Fe, New Mexico, USA

www.greenlion.com

Green Lion Press books are printed on fine quality acid-free paper of high opacity. Both softbound and clothbound editions feature bindings sewn in signatures. Sewn signature binding allows our books to open securely and lie flat. Pages do not loosen or fall out and bindings do not split under heavy use by students and researchers. Clothbound editions meet the guidelines for permanence and durability of the Committee on Production Guidelines for Book Longevity of the Council on Library Resources. The paper used in all Green Lion Press books meets the minimum requirements of American National Standard for Information Sciences—Permanence of Paper for Printed Library Materials, ANSI Z39.48-1984.

Set in 12-point Arno Pro.
Printed and bound by Sheridan Books, Chelsea, Michigan.

Cover design by William H. Donahue.

Cataloguing-in-publication data:
Claudius Ptolemy
The Almagest: Introduction to the Mathematics of the Heavens /
Selections translated by Bruce M. Perry
Edited with notes by William H. Donahue

Includes index, bibliography, biographical notes

ISBN-13: 978-1-888009-43-9 (sewn softcover binding)

1. Ptolemy, Claudius. 2. History of Astronomy
3. History of science. I. Bruce M. Perry (1951–). II. William H. Donahue (1943–). III. Title

Library of Congress Control Number: 2014952353

Cover illustration adapted from Raphael, *School of Athens,* by W. H. Donahue. The crowned figure in the foregound has been identified as Ptolemy.

Contents

The Almagest ~ Book X

Preliminaries to Book XI

The Green Lion's Preface

The Green Lion is gratified to see this book finally reaching publication. Its gestation stretches back to the Lion's earliest years: we first announced a Green Lion Ptolemy Reader in our newsletter of 1997. We wanted an edition of *Almagest* that could be used in one-semester college courses. Some of our history of science colleagues were teaching such courses, and we thought that, with a suitable selection, we could encourage more to do so. We talked both to colleagues who were already teaching such courses and to others who might do so if they had an inexpensive, suitable text, asking what material they would want to have in such a Reader.

At that point the *Almagest* was out of print, except for the version included in the *Great Books of the Western World*; and it, too, soon became unavailable. So the Green Lion commissioned Bruce Perry, of St. John's College in Santa Fe, to undertake a new translation from the Greek of the selections we proposed to include.

In 1998 the Green Lion prepared a grant proposal, submitted to the National Endowment for the Humanities under the auspices of St. John's College. We proposed to bring together a group of faculty from the Santa Fe campus of St. John's College and a group of faculty from other institutions with special expertise in teaching history of science in the context of the humanities, to prepare a new set of source materials in ancient astronomy, and especially Ptolemy, for use in undergraduate and graduate classes. The resulting materials were to be in a variety of media, both printed and electronic, and were intended to lead students to a better understanding of how science evolved, how it works, and how it is related to other branches of learning.

Although the grant was not awarded, we proceeded with the Green Lion project. Bruce Perry completed a draft of the translation, which saw several years of use in freshman and sophomore mathematics tutorials at St. John's College. Each year the faculty teaching the tutorials submitted comments and suggestions for revisions and clarifications, and Perry continued to refine the translation.

In the summer of 2009 a faculty study group at St. John's undertook a review of the Almagest selections and created a set of notes and study materials.

In 2013 William H. Donahue reviewed the entire translation and further revisions were made. Then Green Lion editors Donahue and Howard Fisher prepared the Preliminaries and notes and created or revised diagrams, incorporating and expanding upon teaching aids that had been previously used at St. John's.

About this edition

This book has been designed by teachers and scholars to provide the functionality, convenience, and beauty that is characteristic of a well-produced printed book. Great care has been taken to ensure that all features of the format assist the study of this challenging text. Diagrams are repeated as necessary so that pages do not have to be turned to refer to them. True footnotes, rather than endnotes, assist the reader with obscure terms or difficult steps in the argument. Ample margins allow for writing of notes and comments and for additional diagrams and sketches.

Ptolemy's mathematical proofs, like most ancient mathematical manuscripts, were written out in continuous lines, as if they were ordinary prose. We have inserted line breaks in these proofs to distinguish the individual steps of the argument, which will make it easier for readers to follow the text.

Specific features

Gray shading beginning at the left margin indicates that the overlying text is written by the editors and is not Ptolemy's text.

Numbers in the inner margin in the format H123 refer to the corresponding page number of the Greek text in Heiberg's edition. As of this writing (2014), the complete text is available for download at www.wilbourhall.org. Note that the text was published in two volumes, the first covering Books I-VI and the second covering the remaining books.

Proposition and Observation numbers are provided by the translator as an aid in following Ptolemy's text, and also as a reference to the translator's step-by-step summaries of Ptolemy's proofs. It is our intention to make these summaries available online for free download: see the Green Lion web site (www.greenlion.com) for current information.

Footnotes are all by the editor or the translator.

Diagrams are drawn by the editors, mostly following the Heiberg text, but with some modification where that appeared desirable. *Illustrations and photographs* are by the editors, who have also added some diagrams to Book I to assist in visualizing Ptolemy's description of the universe.

Square brackets [ ] indicate references to external texts such as Euclid's *Elements* or Euclid's *Data.* They are also used to enclose notes that describe omissions.

Angle brackets ⟨ ⟩ indicate interpolations in the text, provided by the translator. These include references to diagrams that are not Ptolemy's but have been provided as aids.

Curly brackets { } are used for cross references within the text, such as citations of other parts of the *Almagest,* Heiberg page numbers, and proposition numbers inserted by the translator.

The Green Lion's acknowledgments

This edition has benefited enormously from the cumulative experience of more than 75 years of studying the *Almagest* with all freshmen and sophomores at St. John's College. Over the years, dozens of faculty have made corrections to translations, contributed notes, and written more detailed explanations. Especially noteworthy is Robert Bart's small fascicle, which helped more than one generation of students through the text. J. Winfree Smith likewise produced commentaries on several aspects of Ptolemy's project. Other notes and comments, usually anonymous, appeared in later years, resulting in a rich but somewhat jumbled collection of ancillary material.

In view of this situation, and in the interest of giving Bruce Perry's draft translation a thorough review, Lynda Myers organized and led a faculty study group in the summer of 2009. Members of the study group included Bruce Perry, Bill Donahue, Michael Wolfe, and Lauren Brubaker. Ms. Myers did an extraordinary job of leading the group through the entire body of selections from the *Almagest* together with the inherited body of College ancillary notes, critiquing the translation thoroughly, identifying the most valuable of the notes, and determining which passages in Ptolemy's text seemed to require additional explanation. The result was a great improvement in both translation and commentary; however, it was Ms. Myers' opinion, at the end of the process, that a new commentary should be written, incorporating the material assembled by the study group, but composed afresh, so as to make up a more coherent whole.

This task was undertaken by Howard J. Fisher, long time Tutor at the College and associate editor at Green Lion Press. His work formed the basis of the Preliminaries to the *Almagest* as a whole, as well as to Books I, II, III, and IX.

As editor of the entire project, William Donahue reviewed the whole translation, consulting the Greek as necessary, and revised Howard Fisher's draft Preliminaries. He also wrote all the annotations to Books X–XIII.

Special thanks is due to James Evans, of the University of Puget Sound, for reviewing the text and suggesting some improvements. In particular, it is due to Prof. Evans that we have included the later chapters of Book XI, which present Ptolemy's method for computing planetary positions. This addition facilitates the generation of authentically Ptolemaic planetary positions for any date. Tables for all five of the classical planets are included.

Finally, we would like to thank St. John's student Max Walukas, who, as Green Lion Press's summer intern, did the initial layout of the text in InDesign.

Dana Densmore
Willilam H. Donahue
for the Green Lion

Translator's Preface

The Greek text used for this translation is Heiberg's; only a few variants are adopted in the translation and some adopted by Toomer are duly noted.

In keeping with my charge from Green Lion Press, this translation's priority is faithfulness to the Greek rather than elegance of English where that requires departing from the original text. It does take some liberties in breaking large sentences into smaller ones, supplying some elliptical omissions, and (in a handful of desperate cases) paraphrasing what seems to be the intended sense.

Ptolemy's Greek is challenging in many ways. He is given to extremely long, paratactic sentences. One of his masterpieces in this regard is the beginning of Book 12: here he presents two complex, equivalent theories of retrogradation at once in a single sentence (H450.8-451.22). At the other extreme, he can be so elliptical in places that it is hard to decide precisely what needs to be supplied. And resolving even his short sentences into clauses is sometimes confusing; there are many stubbornly ambiguous cases. In general, there is a sort of opacity, even awkwardness, to Ptolemy's writing, especially when he is providing a larger frame for a topic or presenting a philosophical discussion.

Where the translation's English reproduces these difficulties so well that the reader is in danger of failing to follow Ptolemy's argument, Green Lion notes offer suggested clarifications.

Footnotes to Ptolemy's text were supplied either by myself or by the editor, William Donahue. References of the form {H100.10} are to Heiberg's page and line numbers; those having the form [Euc. 1.1] are to book and proposition numbers in Euclid's *Elements*. Proposition, Observation, and Calculation numbers have been supplied by the translator for ease of identification.

Translator's Acknowledgments

Many of my St. John's colleagues bravely used the first draft of this translation in their tutorials on the Santa Fe and Annapolis campuses. Many offered helpful criticisms and suggestions; and students were generous with their comments as well. Given the passage of time and numerous rewritings, it is impossible properly to acknowledge everyone. A short list includes, among my Annapolis colleagues, Jeffrey Black and Brendon Lasell; and in Santa Fe Frank Hunt, Michael Rawn, Lauren Brubaker, and Michael Wolfe. Lynda Myers was extremely helpful in improving the translation in many ways. I would especially like to thank William Donahue for his careful checking of the entire translation.

Bruce M. Perry
Santa Fe, New Mexico
August 2014

PRELIMINARIES TO *THE ALMAGEST*

Ptolemy and his readers will have been familiar with the following phenomena (φαίνω, appear), along with certain fundamental concepts that are directly based on them. For convenience, we will describe and illustrate the phenomena as they present themselves in the northern hemisphere.

Appearances of the Fixed Stars

1. *Diurnal motion of the fixed stars.* In common parlance, the adjective *diurnal* (from Latin *dies*, day) refers to daily occurrences, as opposed to nocturnal ones; but in astronomical discourse it refers to the cycle consisting of day and night together.

The "fixed" stars preserve invariable positions relative to one another in the heavens; all other heavenly bodies move about in relation to them. Groups of fixed stars have traditionally been regarded as forming patterns—the so-called *constellations.* Here is a sketch of the constellation Ursa Major (Big Bear). It contains the stars of the Big Dipper, whose handle forms the bear's neck, head and nose.

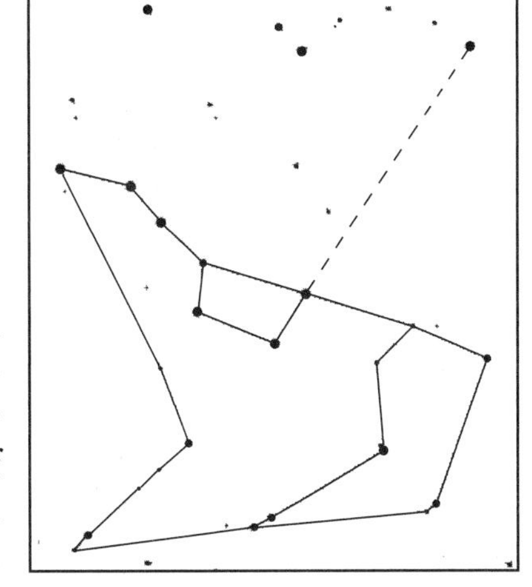

Notice that two of the stars in the Dipper point to a prominent star located roughly above the bear's haunch. Although that star is not part of the constellation, it has a special significance, as we will see shortly.

The stars become visible after the sun sets. During the course of one night the stars describe what seem to be circular arcs in the sky. The circular character of these paths is more evident in certain regions of the sky than it is in others.

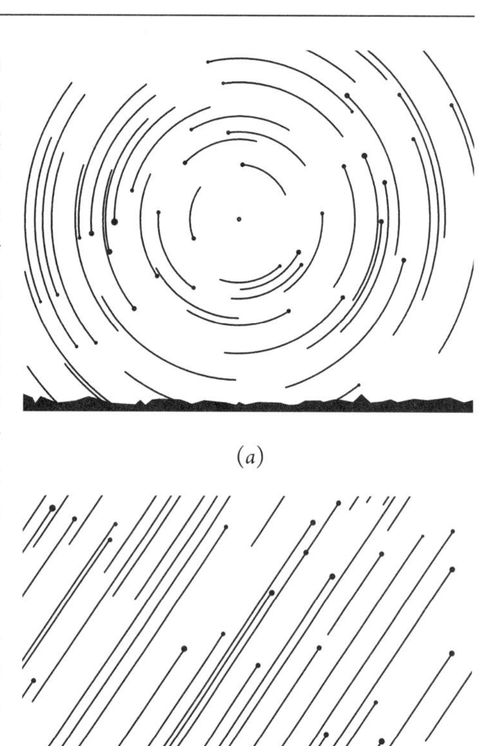
(*a*)

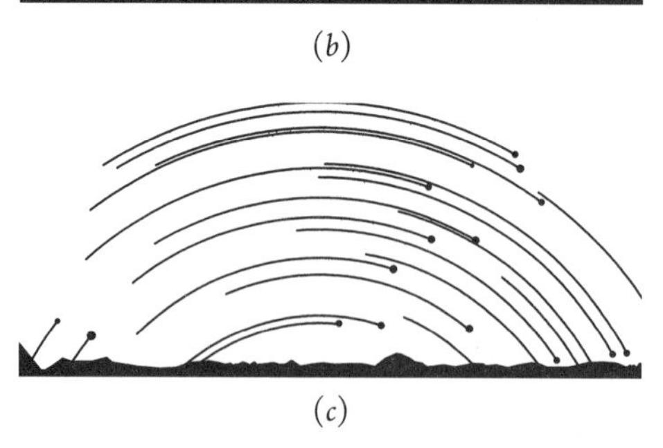
(*b*)

(*c*)

Here are three views of the sky; the filled circles represent bright stars, while the solid arcs and lines depict their "trails" over the previous four-hour period. Views (a) and (c) are in opposite directions, while view (b) is midway between them. The apparent intersection of earth and sky in each of these views is called the *horizon*. In the following discussion, we shall assume that any features on the horizon are distant enough for the horizon to be considered flat.

2. *North and south.* View (*a*) suggests that there is one point in the heavens about which all the stars appear to revolve. One star actually seems to occupy this point, or *pole* of revolution: it is the very star we previously noted in relation to Ursa Major and the Big Dipper. Because of its apparent location at the pole it has been named Polaris (from Latin *stella polaris,* polar star). Careful observation, however, reveals that Polaris actually moves in a small circle about the central point, and thus does not precisely coincide with the pole.

We call the central point in view (*a*) the *celestial north pole*; and the direction on the earth's surface from the observer towards the celestial north pole is called the *northern* direction or, simply, *north*. The opposite direction is *south*. Can we infer from view (*c*) that there must also be a south celestial pole, even though it falls below the horizon and is therefore not visible?

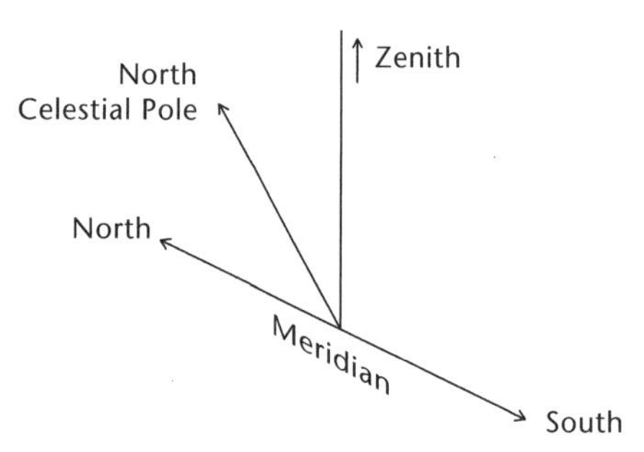

3. *The zenith and the meridian.* At any place on earth, the location directly overhead is called the *zenith,* that directly beneath is called the *nadir*. A line running north and south on the earth's surface is a meridian line, and the plane containing the meridian line and

the zenith is the *plane of the meridian* for the location in question. It should be clear that the plane of the meridian for any geographical location must also contain the north and south celestial poles.

4. *East and west.* Although the stars provide a clear phenomenal basis for defining north and south, it is hard to imagine what stellar appearances would suggest the perpendicular directions, east and west. Etymologically, moreover, the probable origin of the word "east" (according to the Oxford English Dictionary) is ultimately an Indo-European base meaning *to become light* (in the morning). Thus, though it may be convenient simply to *define* east and west as being perpendicular to north and south, we may hope to find a deeper basis for the perpendicular direction when we consider solar phenomena.

5. *Stars that never rise and set.* View (*a*), towards the north celestial pole, shows that stars lying sufficiently close to the pole will never intersect the horizon and thus will never rise or set. It is true that they become invisible during the day, but that is only because they are overwhelmed by the light of the sun, not because they are hidden from view. When Ptolemy speaks of stars that are "ever-visible," (Book I, Chapter 3), he means the stars that never rise or set. Modern terminology calls these stars "circumpolar." The word seems inapt, since *all* stars circle the pole; nevertheless it has become customary.

For the same reason that some stars near the north celestial pole are ever-visible in Ptolemy's sense, we can infer that some stars in the southern part of the sky (view *c*) are *never* visible. But as we shall see, the relation of the horizon to the sky depends on our geographical location; and whether a given star is *circumpolar, rises and sets,* or is *never visible,* depends on where we are on the earth with respect to north and south.

6. *The "spheres."* If our geographical location is such that some stars rise and set, but not all do, the arcs described by the rising and setting stars will intersect the horizon at oblique angles; this will be true in whatever direction we look but is most evident in views such as (*b*). The aspect of the heavens present to us at any such location is what Ptolemy calls the *oblique sphere.*[1] An observer at such a location is said to be "in" the oblique sphere.

If we travel northward from a location in the oblique sphere, the north celestial pole rises in the sky; more stars appear circumpolar and fewer rise and set. If we continue to travel in the northerly direction we will eventually reach a point where the celestial pole is directly overhead. *All* the stars are then circumpolar and the arcs described by them are parallel to the horizon. This aspect of the heavens is called the *parallel sphere.*

If, on the other hand, we travel southward, the north celestial pole descends towards the horizon. Fewer stars are circumpolar and more of them

[1] That the aspect is *oblique* we have explained already; why it is a *sphere* is a topic Ptolemy will discuss in Book I, Chapter 3.

rise and set. If we continue to travel in the southerly direction we will eventually reach a point where *both celestial poles* lie in the horizon, *all* stars rise and set, and the arcs that the stars describe intersect the horizon at right angles. Ptolemy calls this aspect of the heavens the *right sphere*. An observer at such a location is said to be "in" the right sphere.

7. *Latitude.* It is clear from the foregoing discussion that the angle of the celestial pole above the horizon can serve as an index of how far north or south we are on the earth's surface. That angle is called the *latitude* of our location. When we are in the right sphere, we are said to be at 0° latitude. We may also be said to be "on the equator," but that phraseology will have more meaning when we discuss solar phenomena. Similarly, when the north celestial pole is directly overhead we will be said to be at 90° north latitude. We may also be said to be "at the (terrestrial) North Pole." Locations between these extremes have latitudes greater than 0° but less than 90° north.

Similarly, geographical locations lying further south than 0° have latitudes between 0° and 90° south. The place on earth having 90° south latitude is the (terrestrial) South Pole.

Appearances of the Sun

Like the stars, the sun describes what at least resembles a circular arc in the heavens between its rising and its setting. But viewing the sun's arc is far more challenging than watching the stars' paths because it is so hazardous to look at the sun directly. Since the sun shines brightly enough to cast a shadow, however, we can infer many characteristics of its motion from the behavior of its shadow.

8. *The gnomon.* The Greek word γνώμων means "one who knows" or, by extension, a judge or indicator. Ptolemy's gnomon is a vertical rod erected on a horizontal surface; it may be used to measure angles above the horizon. In this photograph, the gnomon's height *AB* is fixed, while the length of its shadow *AC* is small when the angle of the sun above the horizon—angle *BCA*—is large; and large when angle *BCA* is small. The angle of the sun (or any celestial object) above the horizon is called its *altitude*.

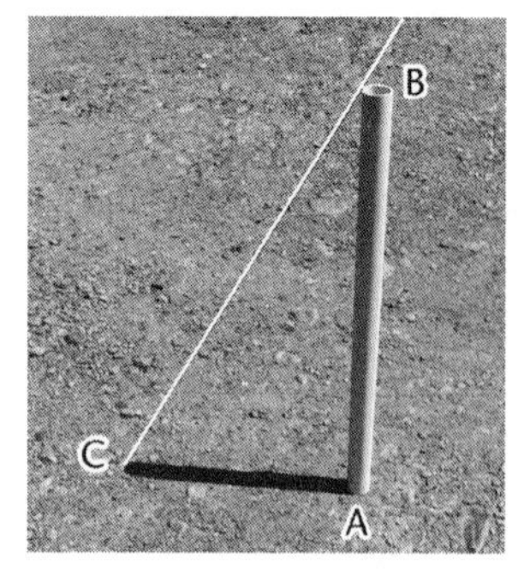

9. *Diurnal motion of the sun.* The sun rises on the horizon in the general direction called *east*; we will define "east" more precisely later. The gnomon's shadow is usually too indistinct to discern either at sunrise or sunset, but within an hour or so after sunrise, as the sun ascends in the heavens, the shadow generally becomes useful.

The diagram shows a typical pattern, with shadows recorded at one-hour intervals. We are facing nearly south. As the sun moves through the sky from east to west, the shadow travels from west to east. Notice that the *shortest* shadow is cast along the north-south line. Clearly, then, the sun reaches its highest altitude of the day when it crosses the plane of the meridian.

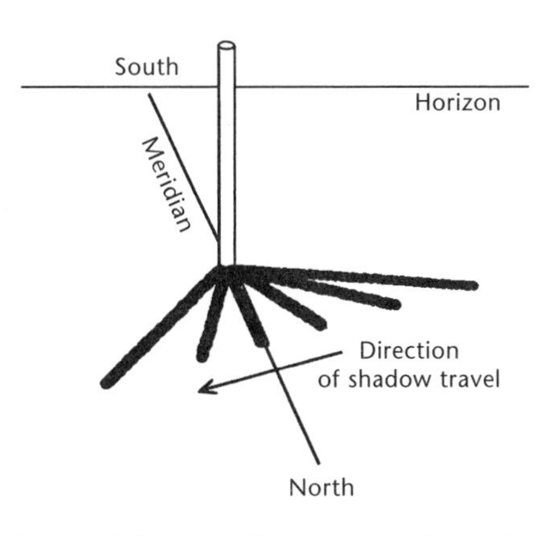

When a celestial body reaches the plane of the meridian it is said to *culminate*. Culmination of the sun constitutes *local noon*.[1]

10. *Annual motion of the sun*. As mentioned earlier, the sun does not maintain a constant position with respect to the fixed stars but moves among them. This motion manifests itself not only with respect to the stars but also with respect to the earth. Let us first consider its relation to the stars.

Of course we cannot see the stars when the sun shines. But shortly after sunset, as the stars begin to emerge, we may notice a given constellation appearing very low on the horizon, and in the same general direction as that along which the sun had set.

This drawing shows the sky about one hour past sunset. The sun is of course not visible, being below the horizon. But having observed how rapidly it had been descending prior to sunset, we can make a rough estimate of where it is now. And if we are familiar enough with the constellation Gemini to recognize it from its upper half (which alone is visible), we see that the sun is just beginning to come into what we call *conjunction* with Gemini.

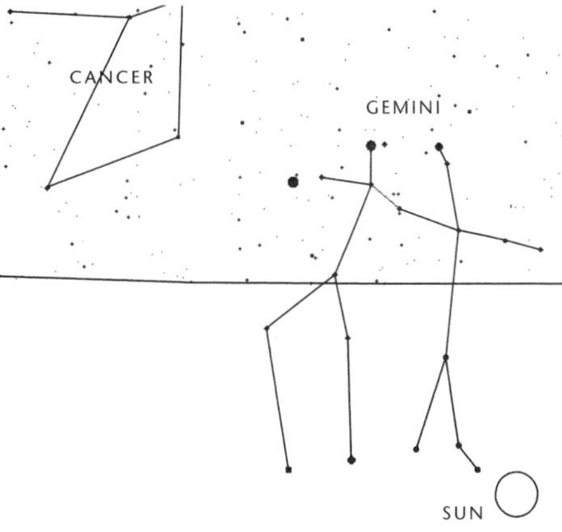

We may also be able to discern some of the stars of the constellation Cancer. If we observe again at sunset about a month later, it will be the constellation Cancer, not Gemini, that appears on the horizon. The sun

[1] Our clocks indicate *civil time*, which represents a compromise among many factors besides what the sun is doing. Therefore local noon as indicated by the gnomon will not necessarily coincide with noon on the clock.

will then be in conjunction with Cancer; it will have moved from one constellation to the next. These conjunctions occur in a cycle, which continually repeats itself. The length of the cycle constitutes, roughly, the *year*. (Ptolemy will advance a more accurate definition of the year in Book III.)

11. *The zodiac.* The sequence of constellations with which the sun is in conjunction during the course of its cycle among the stars is called the zodiac. The zodiac is a wide band of stars, broad enough to contain not only the path of the sun's motion but, as we shall see later, the courses of the moon and the five planets as well.

Traditionally, twelve constellations make up the zodiac; and if we represent the sun's cycle of conjunctions as a circle, each constellation will be associated with one twelfth of the circle, or 30°. The following chart shows the order of the divisions of the zodiac.

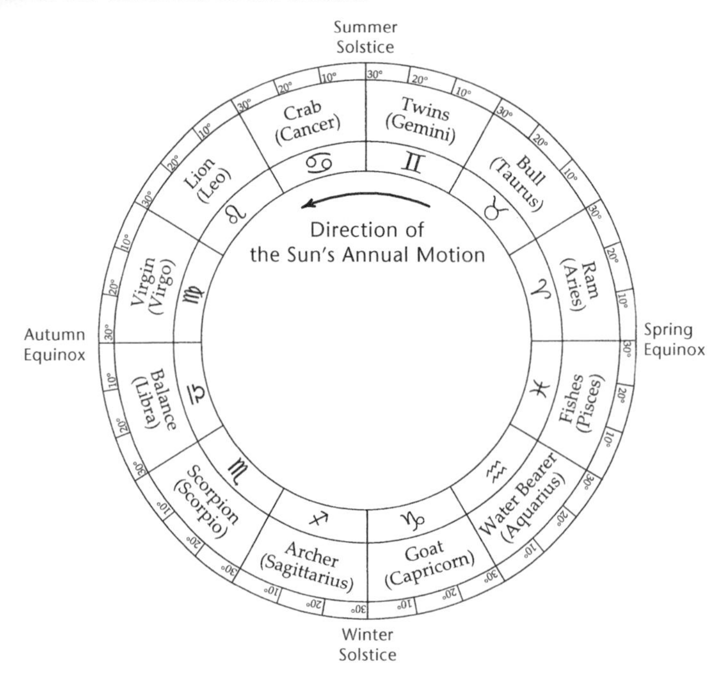

12. *The solstices and equinoxes.* The diurnal arcs described by the fixed stars are, essentially, fixed; if we mark the point on the horizon at which a given star rises or sets on a particular night, it will be found to rise or set at the same point (but not at the same time) weeks, years, and even decades later. Not so with the sun; for most of the year the sun never rises or sets in the same place on two successive days. Instead, from about late June until late December, its rising and setting points move continually southward on the horizon; from early January until mid June they move northward again, in a cycle that appears to have the same length as the sun's cycle through the zodiac.

For example, if one's western horizon has mountains or other features that can be used as landmarks, a sketch or other record of the sun's setting positions over a number of weeks—a "horizon calendar"—will make its north and south motion dramatically evident. If observation is continued throughout a whole cycle, we find that the sun's setting position oscillates between two extreme positions on the horizon, one towards the north and one towards the south. When the sun reaches either of these positions, it appears to stand still for two or three days; it then reverses direction and heads toward the other extreme. These occasions are called *solstices* (Latin *sol,* sun + *sistere,* to stand still); alternatively *tropics* (Greek τροπή, turning). They are indicated in the drawing.[1]

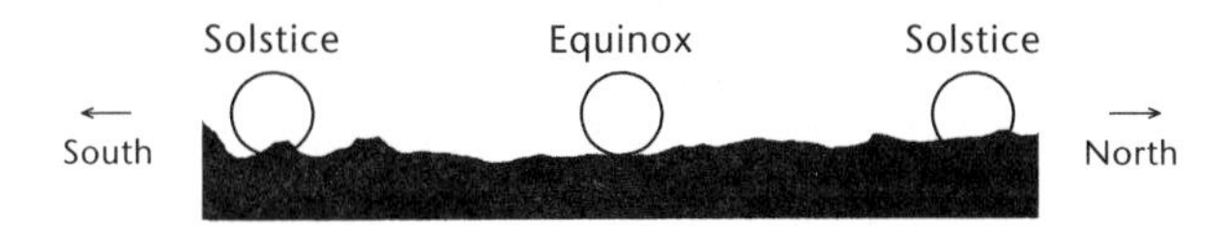

Midway between the solstices lies the position of *equinox* (Latin *aequus,* equal + *nox,* night) so called because when the sun sets in that position, day and night will be of equal duration. In the oblique sphere, whenever the sun sets in a position south of the equinox, the night is longer than the day; while whenever it sets in a position north of equinox, the day is longer than the night. In Book III Ptolemy will offer a reason why there is a correlation between the sun's position and the length of the day.[2]

The same solar motion is manifest by the noon shadow of the gnomon. At the time of summer solstice, when the sun rises and sets at its northernmost extreme, it is also observed to produce the shortest of all the noon shadows—which indicates that it achieves its maximum altitude on the meridian at summer solstice. Similarly it produces the longest of all noon shadows on the day of the winter solstice—which means that the sun has minimum altitude on the meridan at winter solstice. At equinox, the sun's noon altitude will be halfway between its highest and lowest altitudes.

Not only does the sun's north and south motion have the same cycle as its motion through the zodiac, the individual solstices and equinoxes subdivide that cycle into four very nearly equal intervals. Traditionally, the cycle about the zodiac is considered to begin with the Spring Equinox.

---

[1] In the northern hemisphere, the northern solstice coincides with the summer season and is in fact called the Summer Solstice; the other extreme position is the Winter Solstice.

[2] Ptolemy's account will also explain why days and nights are *always* equal in the right sphere and why, at the poles, the sun is continuously visible for half a year and then not visible at all for another half year.

13. *More on east and west.* Even though the position of sunrise is almost always varying, it has been sufficient for our purposes up to now to define "east" as the direction of sunrise. But the discovery of equinox distinguishes one "eastern" direction from all others. Moreover, if we draw lines to the point of sunrise and sunset at the time of equinox, those lines will coincide and will be *perpendicular to the meridian line.* Is this a satisfactory rationale for defining east and west more precisely, as the directions perpendicular to north and south?

North, South, East, and West are our *cardinal* directions, the directions of paramount importance. It is interesting to note that archaeoastronomy has discovered ancient cultures in both central and north America whose cardinal directions seem *not* to have been mutually perpendicular.

Observing Instruments

14. *Armillary sphere.* The armillary sphere is an observing device made up of interconnected rotating rings, or *armillae.* By suitably orienting the rings and their axes, we can emulate the motion of any desired celestial body; then, by sighting across the inner ring at the appropriate degree mark, we can observe the body in the heavens.

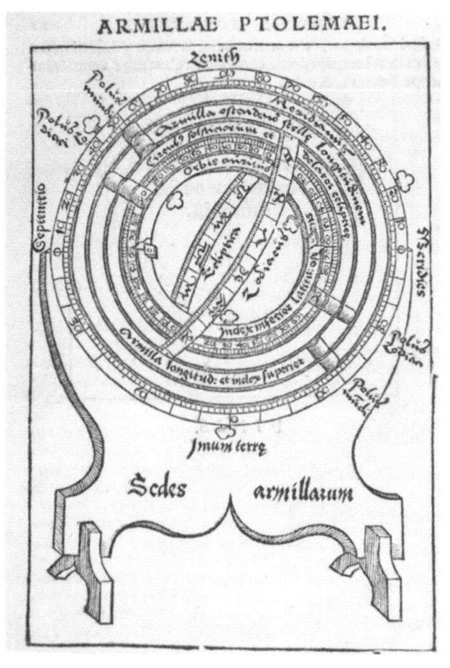

Using the instrument in this way assumes that angles about its center exactly correspond to angles about the center of the celestial sphere itself—or, in other words, that the center of the armillary is located at the center of the celestial sphere. Since we know that this is not in fact the case (for the center of an armillary is actually located on the earth's surface), the successful use of armillaries implies that the radius of the earth must be negligibly small in comparison to the radius of the celestial sphere. Ptolemy will draw that very conclusion in Book I, Chapter 6.

15. *Gnomon.* In Section 8 of these Preliminaries we discussed the gnomon's shadow only in respect of its length and its direction. But the ratio of the gnomon's height to the length of its shadow enables us to calculate the sun's altitude—its angle above the horizon; see the discussion of the *tangent* in Appendix 1 on Trigonometric Functions.

Earlier we noted the special importance of the Sun's altitude at noon. The following two instruments are designed specifically to obtain noontime altitudes.

16. *Quadrant.* In Book I, Chapter 12, Ptolemy describes "a stone or wood plinth," one face of which is aligned with the plane of the meridian and inscribed with a quadrant, or 90-degree arc. The instrument shown in this photograph resides on the campus of St. Mary's College, Moraga, California. In this design, the shadow cast by a horizontal peg indicates the noon angle of the sun directly.

17. *Meridian Ring.* Ptolemy refers to this instrument as "a ring of bronze" in Book I, Chapter 12. Like the quadrant, it is set up in the plane of the meridian, that is, vertically parallel to the north-south line. Thus the east-west line is perpendicular to the plane of the ring and parallel to the horizon.

Pictured below is a modern instrument that combines the bronze ring and the stone plinth. The length of the shadows suggests that the photograph was taken about an hour before noon. At a time closer to noon, the pivoting bar can be rotated until the notch on the upper index casts its shadow on a

line scored on the lower index. The angle of the sun above the horizon may then be read on the calibrated ring. The device shown in the photograph is used at St. John's College, Santa Fe, New Mexico.

The Sexagesimal System

18. The number system used by Ptolemy differs from the modern decimal system. The difference does not affect whole numbers, but when expressing fractions of whole numbers he uses the *sexagesimal system,* which is based on multiples of 60, rather than the decimal system, which is based on multiples

of ten. The sexagesimal system is not wholly unfamiliar to us, for we still use it to express fractions of hours and of degrees; but Ptolemy carries it to a level of precision that is no longer common today.

When we write the decimal 24.52, we know it means 20 + 4 + 5/10 + 2/100. In sexagesimal notation, this same number would be expressed as 24;31,12 parts—that is, 20 + 4 + 31/60 + 12/3600 parts or units. (An alternative notation is $24^{p}\,31'12''$.) Just as the place of a number after the decimal point indicates the degree of subdivision of the unit into 10, so the place of a number after the semicolon indicates the degree of subdivision of the unit into 60. [1]

In the example just considered, the first position after the whole number indicates a fraction with denominator 60, while the number 31 is the number of "minutes" or sixtieths. In the next position, the denominator is 60^2, and the number 12 is the number of "seconds," and so forth. Thus 24;31,12 parts is read "twenty four parts, thirty one minutes, twelve seconds."

We may use as many sexagesimal places as the matter in hand requires or permits. At one point, for instance, Ptolemy gives the sun's mean daily movement correct to six sexagesimal places. In most cases we may be content with numbers expressed to no more than seconds. (Question: How many decimal places does it take to express 1 second?)

Simple addition and subtraction in sexagesimals is perfectly straightforward. For more complicated calculations, a pocket calculator that converts entries to and from degrees, minutes, and seconds will prove handy. The first such calculation Ptolemy reports is the square root of 4500. A modern calculator gives this as 67.082039; then the push of a button converts it to degrees-minutes-seconds, yielding $67°4'\,55.34$. This corresponds to 67;4,55 parts in Ptolemy's notation. (The remainder beyond seconds is left in decimal form on most calculators. Here 0.34 seconds would be equal to $0.34 \div (60{\times}60)$ or .0000944 parts. It will be lost if we round off to seconds.)

For another example, Ptolemy takes the square of 37;4,55 in Book I, Chapter 10. To do this on the calculator we must first convert to decimals. Unfortunately, different brands of calculators have different routines for conversion; here is the procedure for the TI 30, a popular and inexpensive model.

[1] The units or fractions represented by the respective sexagesimal places have the following names in Ptolemy's usage. The chief unit is either the *part* (Latin *pars*)—one of the 120 parts of a circle's diameter—or the *degree* of an angle or arc, or the *hour* of time. Each of these is divided into 60 "small parts" (*partes minutae*, our "minutes"); while each of these, in turn, is divided into 60 "secondary small parts" (*partes minutae secundae*, our "seconds"), and so on.

Enter 37;4,55 as 37.0455. Press the yellow "2nd" button, then the "+" button; this converts from degrees-minutes-seconds to decimal degrees. The result is 37.08194444; square this decimal value, giving 1375.0706, and finally convert back to degrees-minutes-seconds to obtain 1375° 4′ 14, or 1375; 4,14 parts, in agreement with Ptolemy. Always remember to convert sexagesimals to decimals before performing any such calculations, and to re-convert at the end.

If your calculator doesn't do the conversions, or if you would prefer to do the computations manually, there are various ways to proceed. Perhaps the simplest is to convert all the numbers to the smallest units used in the computation (this is a method later adopted by Kepler).

For example, in X.7, H328, Ptolemy calculates the square of $25;58^{p}$. Multiply 25×60 = 1500, which is the number of minutes in 25 parts, then add the remaining 58 minutes. Now all we need to do is square 1558, which is 2427364. Note that since we have multiplied 60ths by 60ths, this number is actually the number of 3600ths, or seconds, in the product.

To convert this back to sexagesimal , we divide by 60, which gives 40456.0666... . The decimal part represents the remaining number of seconds, which we can ignore. We then divide the integer part, 40456, by 60, to get 674.26666.... 674 is the number of parts, and the number of minutes left over is 0.2666×60, or 16. Thus, the square is $674;16^{p}$.

Multiplication of numbers expressed in minutes and seconds is more complex, and is left as a challenge for those who enjoy such things.

Chronology

19. *The Egyptian Calendar*. The Egyptian year is divided into twelve months of thirty days each. Five intercalary days are added each year. The names of the months are as follows:

Thoth	Θώθ
Phaöphi	Φαωφί
Athur	Ἀθύρ
Choïak	Χοϊάκ
Tubi	Τυβί
Mecheir	Μεχείρ
Phamenoth	Φαμενώθ
Pharmouthi	Φαρμουθί
Pachon	Παχών
Paüni	Παϋνί
Epiphi	Επιφί
Mesore	Μεσορή
5 intercalary days	ε ἐπαγόμεναι ἡμέραι

20. *Chronological Table of Kings.*[1]

	Years	Totals
Kings of the Assyrians and the Medes		
Nabonassar[2]	14	14
Nadius	2	16
Chinzer and Porus	5	21
Iloulaius	5	26
Mardokempad	12	38
Arkean	5	43
First Interregnum	2	45
Bilib	3	48
Aparanad	6	54
Rhegebel	1	55
Mesesimordak	4	59
Second Interregnum	8	67
Asaradin	13	80
Saosdouchin	20	100
Kinelanadan	22	122
Nabopoulassar	21	143
Nabokolassar	43	186
Illoaroudam	2	188
Nerigasolassar	4	192
Nabonadius	17	209
Persian Kings		
Cyrus	9	218
Cambyses	8	226
Darius I	36	262
Xerxes	21	283
Artaxerxes I	41	324
Darius II	19	343
Artaxerxes II	46	389
Ochus	21	410
Arogus	2	412
Darius III	4	416
Alexander of Macedonia	8	424

[1] Adapted from Halma's edition of the *Almagest* and F. K. Grinzel, *Handbuch der mathematischen und Technischen Chronologie* (3 vols., Leipzig, 1906). Translated by R. Catesby Taliferro in *Great Books of the Western World*, vol. 16 (Chicago, Encyclopædia Brittanica, 1952).

[2] Nabonassar I (king of Babylon): His era began Thoth 1, equivalent to 26 February 747 BCE. Ptolemy determines the solar "epoch" as beginning at noon, Thoth 1, Nabonassar I. The mean sun was then at 0;45° Pisces, the apparent sun was at 3;8° (see the end of Book III, Chapter 9 below). The Julian Day equivalent of Ptolemy's solar epoch is JD 1448638.

In the present table, a year is counted from Thoth 1 preceding the beginning of each king's reign. Kings ruling less than a year are not listed.

	Years	From death of Alexander	Totals
Macedonian Kings			
Philip	7	7	431
Alexander II	12	19	443
Ptolemy Lagus	20	39	463
Philadelphus	38	77	501
Euergetes I	25	102	526
Philopator	17	119	543
Epiphanes	24	143	567
Philometor	35	178	602
Euergetes II	29	207	631
Soter	36	243	667
Dionysis the Younger	29	276	696
Cleopatra	22	294	718
Roman Emperors			
Augustus	43	337	761
Tiberius	22	359	783
Gaius (Caligula)	4	363	787
Claudius	14	377	801
Nero	14	391	815
Vespasian	10	401	825
Titus	3	304	828
Domitian	15	419	843
Nerva	1	420	844
Trajan	19	439	863
Hadrian	21	460	884
Alelius-Antonine	23	483	907

21. *Other Persons of Interest* (chronological order).

Meton of Athens (Greek mathematician, astronomer and geometer): Flourished in the 5th century BCE. He introduced the 19-year Metonic cycle for synchronous periods of sun and moon, based in part upon an observation he recorded in 432 BCE.[1]

Callippus (Greek astronomer and mathematician): Born about 370 BCE, died about 300 BCE. He measured the inequality of the seasons, which implies irregularity in the sun's motion. He also replaced the Metonic cycle of 19 years for synchronous periods of sun and moon with the Callippic cycle, which is a 76-year period consisting of 4 periods of 19 years each, minus one day. The first Calippic cycle begins 330 BCE.[1]

Euclid (Greek mathematician): He was active in Alexandria during the reign of Ptolemy I (323–283 BCE). His work *Elements*, treating both

[1] For more on the Metonic and Callippic Cycles, see Evans, *The History and Practice of Ancient Astronomy*, pp. 184–187.

plane and solid geometry and number theory, has been characterized as one of the most influential works in the history of mathematics.

Apollonius of Perga (Greek geometer and astronomer): Born about 262 BCE, died about 190 BCE. He is best known for his *Conics,* a treatise on the conic sections.

Hipparchus (Greek astronomer, geographer, and mathematician): Born about 190 BCE, died about 120 BCE.

Archimedes (Greek mathematician, physicist, inventor, astronomer): Born about 287 BCE, died about 212 BCE.

Menelaus of Alexandria (Greek mathematician and astronomer): Born about 70 CE, died 140 CE. His only surviving book is *Sphaerica,* in an Arabic translation. It introduces the concept of the spherical triangle and proves what are now known as the Menelaus Theorems for both plane and spherical triangles.

Claudius Ptolemy (author of *The Almagest*): Born about 85 CE in Egypt, died about 165 CE in Alexandria, Egypt.

Theon of Alexandria: Born about 335 CE, died about 405 CE. He edited and arranged Euclid's *Elements* and wrote commentaries on works by Euclid and Ptolemy. His daughter Hypatia was a famed mathematician.

EPITOME OF THE PTOLEMAIC SYSTEM

The following is a summary of the system of the world constructed by Ptolemy in Books I and III. Some readers will find it useful as an advance overview of Ptolemy's hypotheses. Others will prefer to pass over it until after having studied Ptolemy's own presentation.

The Layout of the World

Figure 1 is a diagram of the world as outlined by Ptolemy in Book I of *The Almagest*. The earth T stands still at the center. About it is an enormous sphere, $AQSP_NR$, the *sphere of the fixed stars,* or celestial sphere. In permanent places on this sphere are located the fixed stars, such as C, which preserve invariable positions relative to one another and so form the constellations. The heavenly bodies are the constellations, the sun, the moon, the five planets (Mercury, Venus, Mars, Jupiter, and Saturn), along with comets and meteors. The earth is spherical, as are all the heavenly bodies, and hardly larger than a point in comparison with the celestial sphere.

The sphere of the fixed stars is in constant uniform rotation about one of its diameters, P_NP_S, called the *axis.* The points on the celestial sphere at the ends of the axis, P_N and P_S, are called the *north* and *south celestial poles.* They

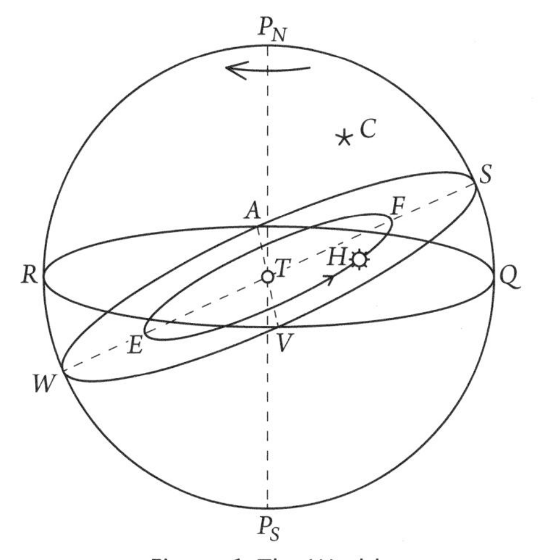

Figure 1. The World.

are distinguished from one another by the difference in the constellations around them. The points where the axis comes through the earth's surface are the north and south terrestrial poles. The direction in which the sphere turns about the axis is said to be from east to west. To an observer above the north pole of the celestial sphere, looking south along the axis, it is the clockwise direction. The time of a complete rotation is about 4 minutes less than a day.

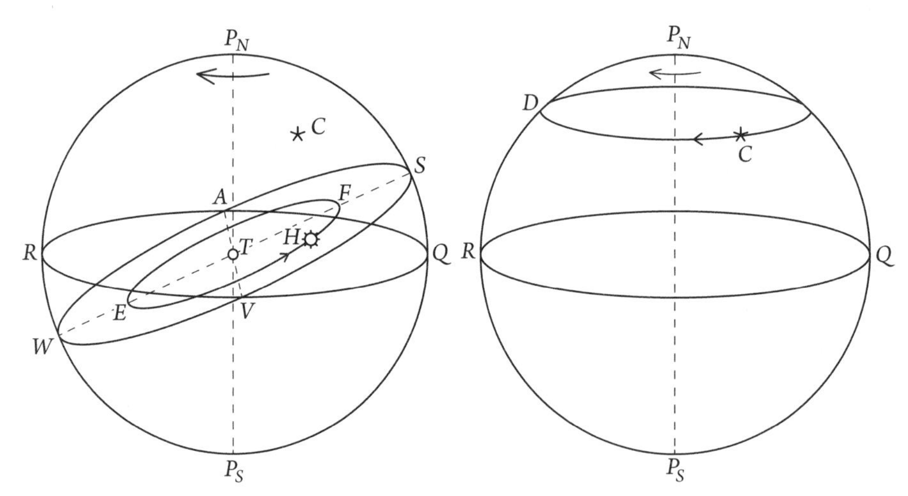

Figure 1. The World. Figure 2. Motion of the Same.

The great circle *QAR* on the sphere, made by passing a plane through the center of the sphere perpendicular to the axis, is called the *celestial equator,* and the corresponding circle on the earth is the *terrestrial equator*. A fixed star such as C is carried by the uniform westward motion of the celestial sphere in a circle having its center on the axis and its plane perpendicular to the axis (Figure 2). All of the other heavenly bodies also partake in this westward diurnal motion. In the *Timaeus,* Plato called this motion "the motion of the same."

While the fixed stars retain their positions relative to one another, the other heavenly bodies (sun, moon and planets) appear to move among the stars. This motion is, in general, eastwards, opposite in direction to and slower than the diurnal motion; Plato called it "the motion of the other." For example, the sun *H* (Figure 1) has, in addition to its daily westward motion parallel to the equator, a slow motion along oblique circle *EHF* from west to east. We will begin by supposing that circle *EHF* is homocentric with the sphere of the fixed stars, and that the sun's motion along it is uniform. These two assumptions are interrelated, however; and in Book III, Chapter 3 Ptolemy will discuss both of them.

Circle *EHF* is much smaller than the sphere of the fixed stars, and its plane is inclined to the plane of the equator. If this circle is projected by lines drawn from the center of the earth upon the sphere of the fixed stars, a great circle *WVS* of the sphere results, intersecting the celestial equator at the angle of about 24 degrees. Circle *WVS* has a fixed position on the sphere of the fixed stars; that is, it passes through a definite series of constellations, the *Zodiac* constellations. The Zodiac is the belt or zone of the celestial sphere bounded by two circles drawn parallel to circle *WVS,* one on each side and each 8° distant from it (Figure 3). The projection of the sun's path, circle *WVS,* is accordingly called the *mid-zodiac circle*; it is also known as the *ecliptic*. The

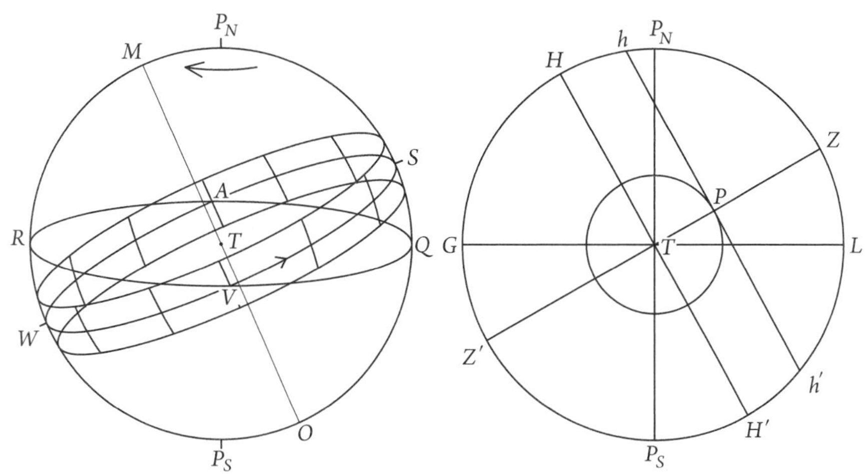

Figure 3. Motion of the Other. The Zodiac. Figure 4. Sensible and Rational Horizons.

Zodiac is divided into twelve equal spherical quadrilaterals, called *signs*, by arcs of great circles drawn perpendicular to the mid-zodiac circle. The length of a sign, that is, the length of the arc of the mid-zodiac circle it contains, is 30°. The sun, the moon, and the five planets, as seen from the earth, are always within the Zodiac.

The axis of the sun's motion through the Zodiac is the line perpendicular to the mid-zodiac circle through its center, with its poles *M* and *O* (Figure 3) lying on the celestial sphere. These poles, together with the poles P_N and P_S of the sphere of the fixed stars, define a great circle of the celestial sphere. The intersections of this great circle and the mid-zodiac circle determine points *S* and *W*, the northernmost and southernmost points of the sun's eastward motion. These are called the *solstices* or *tropic points*, *S* being the summer solstice and *W* the winter solstice. The point *V* where the sun enters the northern hemisphere is called the *spring equinox*; *A* is called the *autumn equinox*. By definition, a *year* is the time the sun takes to complete a circuit of the mid-zodiac circle, defined by a return to the same equinox or solstice. It is assumed that all years are identical in length, and one of Ptolemy's first tasks (Book III, Chapter 1) will be to calculate this length, for which he finds a value slightly less than $365\frac{1}{4}$ days. (Ptolemy will actually express this number to a precision of milliseconds!) Our calendar is so adjusted that March 21st always comes within a day or so of the time when the sun is at the spring equinox.

The Horizon

The plane drawn tangent to the surface of the earth at a place *P, as hh′* (Figure 4), is called the *plane of the sensible horizon* at *P*. The parallel plane through the center of the earth, as *HH′*, is called the *plane of the rational horizon* at *P*. The distance between these planes is the earth's radius. They

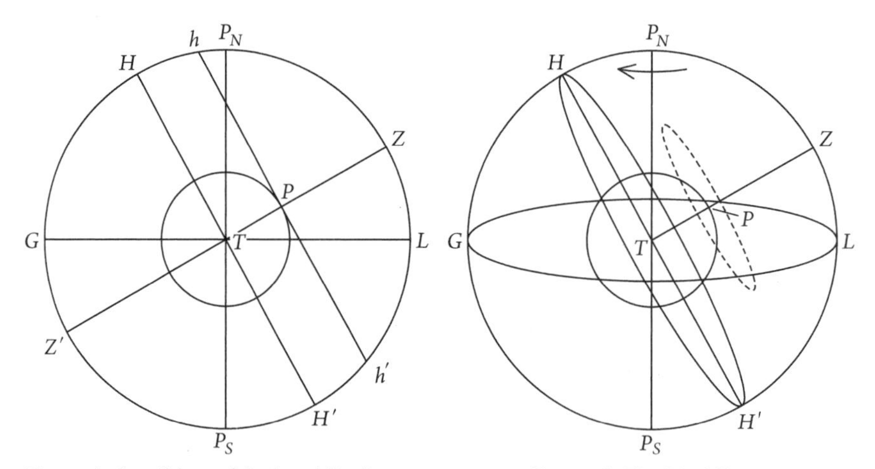

Figure 4. Sensible and Rational Horizons.

Figure 5. The Meridian.

cut the sphere of the fixed stars in two circles known as the sensible and the rational horizons of *P*, respectively. However, the earth's radius is so small in comparison to the radius of the celestial sphere, that two points on the sphere distant from one another by the earth's radius seem to our sight to be one. Consequently we treat the sensible and rational horizons as a single great circle, called simply the *horizon*. The horizon HH' divides that hemisphere of the celestial sphere that is visible to an observer at *P* at any one moment from that hemisphere which the interposition of the earth renders invisible.

Meridian

Given a place *P* on the surface of the earth not at one of the poles, the *plane of the meridian* for that place is the plane which passes through the axis of the celestial sphere and that place. The great circle in which the plane of the meridian cuts the sphere of the fixed stars is called the *meridian* of the place.

The plane of the meridian and the plane of the horizon of a place are at right angles to one another. The *zenith Z* of the place *P* is the point vertically above the observer, in the plane of the meridian and perpendicular to the plane of the horizon. The *latitude* of *P* is the angle between the horizon of *P* and the celestial North Pole—angle HTP_N in Figure 4 or HPP_N in Figure 5. This angle is equal to the angle between the zenith of *P* and the celestial equator—angle *ZTL* in Figure 4 or *ZPL* in Figure 5.

In Figure 6, the great circle *NESW* is the horizon of *P*, NP_NZLSP_S is the meridian of *P*, and *ELWG* is the celestial equator. The whole celestial sphere rotates daily around the axis P_NP_S, carrying with it not only the fixed stars but the oblique plane on which lies the circle traversed annually by the sun. When a star crosses the meridian above any place *P*, it is said to *culminate*. The daily culmination of the sun defines *local noon*.

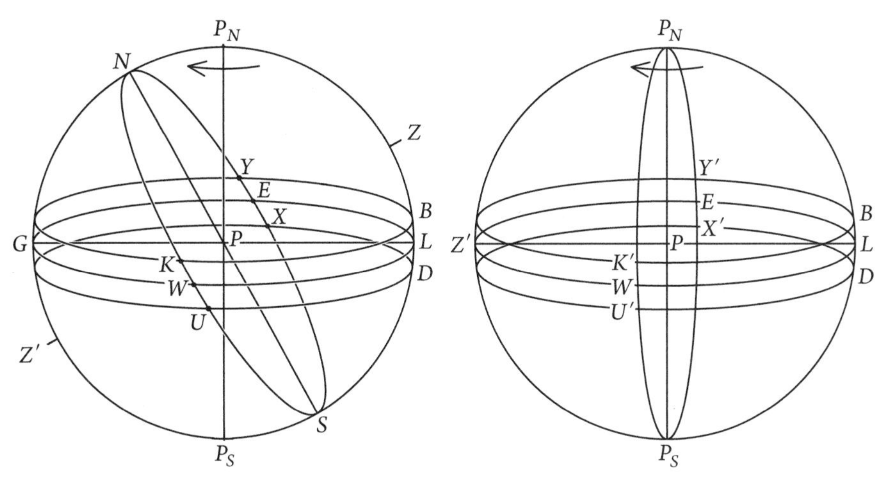

Figure 6. The Oblique Sphere.

Figure 7. The Right Sphere.

On the equinoxes, the sun rises at *E*, comes to the meridian (or culminates) at *L* (local noon for the place *P*), and sets at *W*. On the summer solstice, it rises at *Y*, culminates at *B* and sets at *K*. On the winter solstice, it rises at *X*, culminates at *D* and sets at *U*. Between the solstices its daily path is on circles between these and nearly parallel to them. From Figure 6 it should be clear what determines whether there are some stars that will never set for an observer at *P* and some stars that will never be visible. It should also be clear why the time from sunrise to sunset on a given day will be different at different latitudes.

The Right Sphere and the Oblique Sphere

The *right sphere* is the term Ptolemy uses to signify that aspect of the celestial sphere that is seen by an observer located on the earth's equator. This observer's horizon contains the North and South celestial poles, and the plane of the equator is at right angles to his horizon. Since in each day the sun travels in a circle that is virtually parallel to the equator, its circuit will always be cut in half by the horizon of the right sphere; thus the days will always equal the nights in length.

In Figure 7, a horizon in the right sphere is shown as the plane P_NEP_SW, perpendicular to the plane of the equator $ELWZ'$. The meridian is P_NBDP_S, the zenith is *L*, and the tropics culminate at *B* and *D*. The sun rising at Y' travels in a circle nearly parallel to the equator, coming to the meridian at *B* and setting at K'. The circle $Y'BK'$ is exactly bisected by the plane of the horizon; therefore the sun will spend half of the day above the horizon, half below. Day will always equal night in length.

By contrast, Figure 6 illustrates what Ptolemy calls the *oblique sphere*. It represents a location *P* north of the equator; the North-South axis is inclined to the plane of the horizon, and the equator is oblique with respect to the

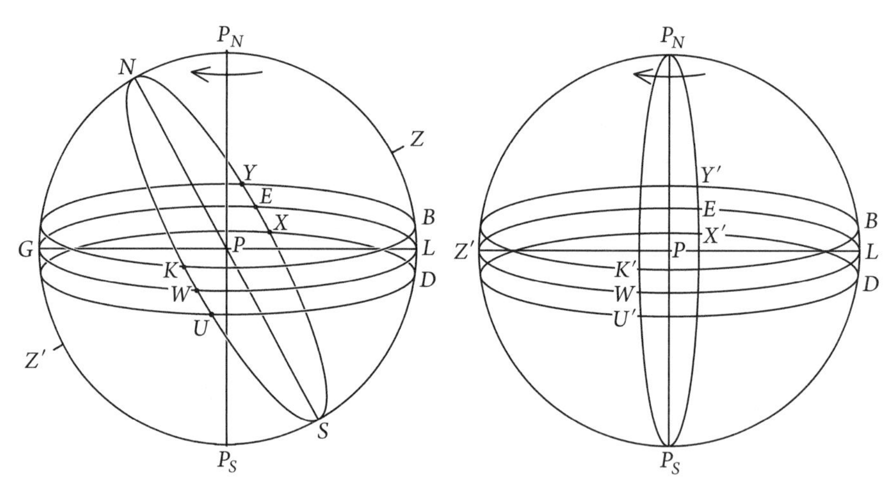

Figure 6. The Oblique Sphere.

Figure 7. The Right Sphere.

horizon instead of being perpendicular to it. Thus when the sun is north of the equator, more than half of any circle it travels in its daily motion will be visible above the horizon, and the days will be longer than the nights. On the other hand, when the sun is south of the equator, less than half of any circle of the sun's daily motion will be visible, and the days will be shorter than the nights. Of all the circles travelled by the sun in its daily motion, only the equator itself is bisected by the horizon of the oblique sphere; therefore equinox occurs when the sun is on the equator.

The oblique sphere is illustrated in Figure 6 by the horizon *NESW* with the same meridian as before, so that both horizons have the same longitude. Since the plane of the horizon is inclined to the plane of the equator, it will not bisect any of the circles parallel to the equator. Thus on the same day that the sun rises at *Y′* in the right sphere, it will rise earlier—at *Y*—in the oblique sphere; and instead of setting at *K′* it will set later, at *K*.

Right Ascension and Longitude

Angular distance measured eastward along the equator, beginning from the spring equinox, is what Ptolemy calls "ascension in the right sphere" or, more simply, *right ascension*. Every meridian intersects the equator at some angular distance from the spring equinox; and that distance (measured eastward) constitutes the right ascension both of that meridian and of every celestial object through which it passes. In Figure 8, *AVC* represents the equator with poles P_N and P_S; *DVF* represents the ecliptic with poles *Q* and *R*. Their intersection *V* is the spring equinox. The meridian P_NPB, drawn through point *P*, determines *VB*, the right ascension of *P*.

Angular distance measured eastward along the ecliptic, on the other hand, is what Ptolemy calls *longitude*. Every celestial object is associated with

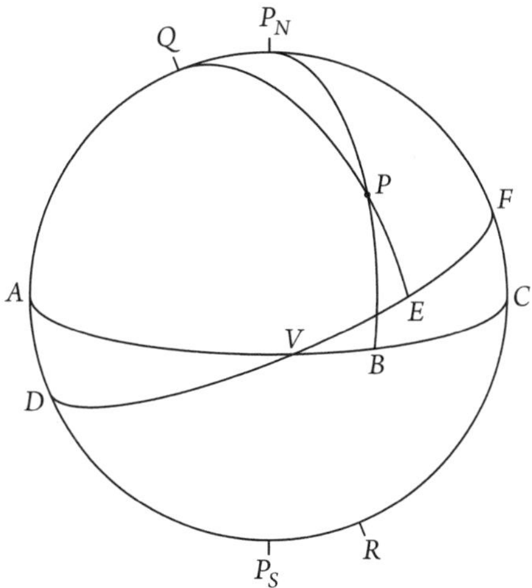

Figure 8. Right Ascension and Longitude.

a longitude by drawing a great circle through that object and the poles of the ecliptic. Every such circle intersects the ecliptic at some angular distance from the spring equinox; that distance (measured eastward) is the longitude both of the great circle and of every celestial object through which it passes. In Figure 8, the great circle *QPE*, drawn both through the poles of the ecliptic and point *P*, determines *VE*, the longitude of point *P*.

Book I

1. Introduction.

H4.7 Those who genuinely engaged in philosophy, Syrus, seem to me quite rightly to have distinguished the theoretical part from the practical part of philosophy.[1] If in fact it turns out that the practical also is at first theoretical, one would nevertheless find a great difference between them. Not only is it the case that many people can possess some of the moral virtues[2] without instruction, but it is impossible to succeed in the study of the heavens[3] without learning: in addition, in practical philosophy the greatest advantage accrues from continuous activity in actual affairs, while in theoretical philosophy H5 from progress in theorems. Hence we thought it proper for ourselves to order our actions amid the presentations of appearances themselves such that even in chance matters we not lose sight of our inquiry into beautiful, well-ordered structure; and to devote our research for the most part towards teaching many beautiful theorems, and especially those properly called mathematical.

In fact, Aristotle quite suitably distinguishes theoretical philosophy itself again into three kinds: the natural, the mathematical, and the theological.[4] Now, all things that are derive their existence from matter, form, and motion. Each of these cannot be observed distinctly and without the others

---

1 Ptolemy has adopted Aristotle's division of philosophy into "theoretical" philosophy and "practical" philosophy. According to Aristotle's scheme (see, for example, *Metaphysics* 1025b3–1026a32), theoretical philosophy, as the name (from Greek θεωρέω, to look at, view as a spectator) suggests, concerns knowledge for its own sake, without an immediate view either to acting or making; practical philosophy concerns knowledge for the sake of doing, making or acting (cf. πράττω, to act or do). "Practical philosophy" includes the arts (medicine, for example), political science and ethics. Aristotle subdivides theoretical philosophy into three types: natural philosophy, mathematics and metaphysics.

2 The moral virtues are within the purview of ethics, a part of practical philosophy.

3 "The study of the heavens" is, according to Aristotle's scheme, a part of mathematics, and thus of theoretical philosophy.

4 Aristotle (*Metaphysics* 1025b3–1026a32) subdivided theoretical philosophy into these three types according to the way the objects each considers are related to matter and motion:

(1) Natural philosophy or "physics" (from φύω, to grow) is concerned with the ever-changing, visible, material world.

(2) Mathematics concerns entities (points, lines, geometric figures, numbers, et al.) that do not exist apart from matter, but may be understood without reference to matter. In Aristotle's view, optics, harmonics and astronomy (as well as arithmetic and geometry) are branches of mathematics.

(3) Metaphysics ("after physics") is concerned with "first causes," with being as such and the properties of being, and with God in so far as the divine nature can be understood by reason alone (natural theology).

in a substance, but can only be conceived of. If anyone should grasp in its pure simplicity the first cause of the first motion of the heavens,[1] he would consider it a god, invisible and unmoved.[2] And he would consider the kind of theoretical philosophy that seeks this out to be theological; such an activity (ἐνέργεια) somewhere on high that concerns what is highest in the heavens could only be conceived of and is absolutely distinct from perceptible substances. The kind of theoretical philosophy that tracks down material and ever-changing quality, concerning white, warm, sweet, soft, and the like,[3] one would call natural (φυσικόν). A nature (οὐσία) of this sort resides among perishable things, for the most part, and below the sphere of the moon.[4] The kind of theoretical philosophy that reveals quality in terms of forms and motions inquires into figure, quantity, and size, and furthermore, place, time, and the like. One would define this as mathematical. A nature (οὐσία) of this sort falls, as it were, between those two.[5] For it not only can be conceived of both with and without perception, but also is a property of all existent things simply, both mortal and immortal. It changes along with things that are always changing relative to their inseparable form, while it preserves unchanged what is unchangeable in form for things that are eternal and of the nature of ether.[6] **H6**

In light of these considerations, we thought that one might call the other two kinds of theoretical philosophy *conjecture* (εἰκασία)[7] rather than scientific knowledge: the theological because it is utterly suprasensible and unobtainable, and the natural because of what is unstable and obscure in

---

1 See Aristotle, *Physics*, Book VIII.5–10 (256a4–267b26) for the argument that the prime mover is an unmoved mover.

2 See, for example, *Metaphysics* 1071b3–1073a13 for Aristotle's argument that the first unmoved mover must be pure actuality, without any potentiality, and thus pure "activity."

3 Compare Plato, *Theaetetus* 171e1–3.

4 According to Aristotle (for example, *On Generation and Corruption* 392b7–331a5) the sublunary world, as contrasted to the heavens, is the realm of change. It is the world of tangible things composed of the four elements, earth, air, fire and water.

5 The objects of natural philosophy are said by Aristotle to be "inseparable" from matter and motion, either in thought or in being—they can neither exist apart from matter and motion nor be understood without reference to matter and motion, because these are essential to their definitions. On the other hand, the eternal, unchanging objects of metaphysics are understood to be "separable" from matter and motion both in thought and in being. The objects that mathematics studies lie between these two in that they are "inseparable" from matter in form and being but are "separable" in thought.

6 See, for example, *De Caelo* 269b16–270b30 for Aristotle's account of the fifth element, ether, which is eternal, ungenerated and indestructible, and exempt from increase or alteration. According to Aristotle, ether is found only in the supralunary realm and is the material basis of the heavenly bodies.

7 *Republic* VI (509d–511e).

matter. For this reason, those who pursue philosophy would never hope to come to agreement about them. But the mathematical kind alone, should anyone attend to it carefully, would provide certain and steadfast knowledge to those who practice it, as its mode of proof arises through indisputable paths, arithmetic and geometry. And we were induced to cultivate as far as possible the whole of this theoretical discipline and especially the part that contemplates the divine and the heavenly. For this alone is concerned with

H7 the inquiry into things that are "always such,"[1] and is thereby itself capable of being "always such" concerning the comprehension it properly produces, which is neither obscure nor irregular (this is a distinctive feature of scientific knowledge). And the mathematical kind contributes to the other kinds of theoretical philosophy no less than they themselves do. For it would most pave the way for the theological kind; it alone is capable of making sound conjectures about the unmoved and separable activity[2] (ἐνέργεια) using the close agreement between the properties (in terms of motions and regularities of motions) of perceptible substances (which produce motion and are in motion) and of substances that are suprasensible and unchangeable.

It would make no trifling contribution to the natural kind of theoretical philosophy. For, roughly speaking, what is in general distinctive of material substance becomes manifest from the specific form of its motion.[3] Thus the perishable itself and the imperishable become manifest from their rectilinear and circular motions, and the heavy and the light, or the passive and the active, from motion towards the center and motion away from the center.[4] And surely it would render men clear-sighted in the nobility of their actions and character because of the uniformity, orderliness, proportion, and modesty it contemplates in the divine. It makes those who pursue it lovers of this divine beauty, and it habituates and, as it were, naturally disposes them to a like structure of soul.

Now we ourselves are striving unremittingly to increase this love for

H8 the study of the "always such." We study what was already grasped in such disciplines by those who pursued them in a genuine spirit of inquiry. And we also propose ourselves to contribute roughly as much as the time between them and us might have produced. And everything that at present we believe to have been discovered we will note down as briefly as possible and in such a

1 Compare Parmenides 28 b 8.29–30 (in Curd, *A Presocratic Reader*, p. 60).

2 See *Physics* VIII.5–10 (256a4–267b26) and *Metaphysics* XII.6–8 (1071b3–1074b14) on the prime mover.

3 See, e.g., *De Caelo* 268b11–270b30 on the characteristic natural motion of each of the five elements.

4 See *De Caelo* I.2–3 (268b11–270b31).

way that those who have already made some progress might be able to follow. In order to make our treatise complete, we will set forth all that is useful for the study of the heavens in its proper order. But so as not to make the account too long, we will merely recount what the ancients precisely determined; what they did not grasp, either at all or in the most useful way possible, we will, as far as possible, elaborate.

2. On the Order of the Theorems.

Now, in the work before us, the first task is to see the general relation of the earth as a whole to the heavens as a whole {I.3–7}. Directly thereafter come the particulars; of these, the first task would be to provide a detailed account of the position of the oblique circle[1] {I.8}, the regions of the earth we inhabit {II.1}, and furthermore the ordered difference[2] among these regions for each horizon according to their latitudes {II.2–13}. For the prior study of these matters makes the investigation into the rest easier. The second task is to provide a detailed account of the motions of the sun {III} and moon {IV–VI}, and what these motions entail. For unless these matters are understood first, it would not be possible to inquire in detail into the stars.[3] Since the account of the stars comes last in terms of the method itself, what relates to the sphere of the so-called fixed stars would reasonably be set forth here first {VII–VIII} and what relates to the five stars called planets[4] would follow {IX–XIII}.

H9

We will attempt to demonstrate each of these matters. We will use manifest appearances and indisputable observations of antiquity and our own time as principles and foundations, as it were, for discovery. And we will apply the ensuing concepts using demonstrations based on geometrical methods.

A preliminary treatment, then, of the general features would be as follows. The heavens are spherical and move as a sphere {I.3}. The earth itself also is sensibly spherical in shape, taken as a whole {I.4}. In terms of position, it is

---

1 *the oblique circle*: The circle traditionally called the *ecliptic*. It will be identified and discussed in Chapter 5 below.

2 *the ordered difference*: Ptolemy does not state in respect of *what* the difference consists; a phrase like "of appearances" is to be understood.

3 *stars*: Here, and frequently in Ptolemy's usage, the term refers not only to the "fixed stars" that constitute the constellations, but also to Mercury, Venus, Mars, Jupiter and Saturn.

4 *planets*: Literally, "wanderers." For Ptolemy, the term refers not only to Mercury, Venus, Mars, Jupiter and Saturn but also to the sun and the moon. It does not refer to comets and meteors, which Ptolemy probably followed Aristotle in regarding as atmospheric phenomena.

situated in the middle of all the heavens, near the center {I.5}. In terms of size and distance, it has the ratio of a point to the sphere of the fixed stars {I.6}. H10 And the earth itself does not move at all {I.7}. We will briefly give an account of each of these things by way of reminder.[1]

3. That the Heavens are Spherical and Move as a Sphere.

It is probable, then, that the ancients got their first ideas about these matters from the following sort of observation. They saw that the sun, moon, and the rest of the stars moved from risings to settings invariably along circles parallel to one another. And at first, they moved upwards from their low point and as if from the earth itself, and gradually rose up to a height. Then again in corresponding fashion they returned and, reaching their low point, proceeded until they completely disappeared, as if they had fallen into the earth. Then again, in turn, after they remained invisible for some time, they rose and set as if anew. And these periods and furthermore the regions of their risings and settings regularly and uniformly corresponded, for the most part.

Their observation of the circular rotation of the ever-visible stars, which revolved around one and the same center, most of all led them to the idea of H11 a sphere. For that point necessarily became a pole of the heavenly sphere. The stars that were nearer to it turned in smaller circles. Those that were farther off made, in proportion to their distance, larger circles in their circuit until their distance reached all the way to the stars in occultation.[2] And of these, those near the ever-visible stars they saw remaining in occultation for a brief time, while those farther off remaining in occultation for a proportionally longer time. So that at first they formed the above-mentioned idea of a sphere through these sorts of observation alone, but soon enough through subsequent study they apprehended as well what followed from them; since all the appearances without exception testify against contrary ideas.

Now consider. If anyone should suppose that the stars move in a straight line indefinitely onward, just as some thought, what manner of explanation could be conceived whereby each will be observed to move every day from the same starting point? For how were the stars that move indefinitely onward able to return? Or how were they not seen as they returned? Or how did they not disappear as their magnitudes gradually diminished? On the contrary,

---

1 *reminder*: Ptolemy assumes that in the next few chapters he will be reviewing arguments' and conclusions that are already familiar to his readers. For an overview of the system of the world whose main features he will now present, see Epitome (pp. 15–21, above).

2 *the stars in occultation*: that is, the stars capable of being below the horizon and therefore being "occulted" (hidden) at certain times.

they are seen to be larger near their own occultations, and are gradually occulted and cut off, as it were, by the surface of the earth.

But the claim[1] that they are kindled from the earth and are again extinguished into it would appear utterly absurd. For let someone concede that all this order in their magnitudes and sizes and, furthermore, their distances, regions, and periods results in so random a way and, as it were, by chance. And let him concede that this whole part of the earth has a kindling, while that part has an extinguishing nature; or, rather, that one and the same part is kindling for some stars but extinguishing for others, and that the same stars actually are already kindled or extinguished for these observers but not yet for those. If anyone, I say, should concede all these things that are so ridiculous, what would we say about the ever-visible stars that neither rise nor set? Or why do the stars that are kindled and extinguished not invariably both rise and set, while those not so affected are invariably above the earth? Surely, for some observers the same stars will not always be kindled and extinguished, but for others will never undergo any of these things. For it is altogether clear that the same stars both rise and set for some observers, but do neither for others. H12

To speak briefly, even if one should suppose some shape other than the spherical for the movement of the heavens, it is necessary that the distances from the earth to the parts of the heavens be unequal, no matter where and how the earth itself is situated. The result is that both the magnitudes and relative distances of the stars ought to appear unequal to the same observers in each revolution, being at one time at a greater distance and at another time at a shorter distance. But this is not seen to occur. And what is more, it is not the shorter distance that makes their magnitudes appear greater at the horizons. Instead, it is the exhalation of the moisture enveloping the earth which comes between our eyes and the stars; just as things thrown into water appear larger, and the larger the deeper they sink. H13

The following considerations add as well to the idea of a sphere. It is not possible for the constructions of time-instruments to concur on any other hypothesis than this one alone. Also, the heavenly bodies have a movement that is unobstructed and most readily accomplished of all; among plane figures the circular has the most readily accomplished motion, while among solid figures it is the sphere. Likewise, among different figures that have an equal perimeter, those with more sides are larger. Hence, the circle is larger than other plane figures, the sphere than other solid figures, and the heavens are larger than all other bodies.

What is more, it is possible to pursue this sort of conception from certain physical considerations. For example, among all bodies ether has H14

[1] See Heraclitus 22 B 30.

the smaller and more uniform elements, while among bodies with uniform elements surfaces with uniform elements have more uniform elements. But the only surfaces with uniform elements are the circular among planar, and the spherical among solid surfaces. And since the ether is not planar but solid, it follows that it is spherical. Similarly, nature constructs all earthly and perishable bodies entirely out of round shapes, albeit without uniform elements. But it constructs all the divine bodies in the ether again out of uniform and spherical shapes; since, if they were planar or disk-shaped, they would not display a circular shape to all who see them at the same time from different regions of the earth. Therefore, it is also reasonable that the ether that encompasses them, being of the same nature, is spherical and, due to the uniformity of its elements, moves circularly and uniformly.

4. That the Earth Also Is Sensibly Spherical as a Whole.

That the earth too is sensibly spherical, considered as a whole, we might best understand in the following way. For again it is not possible for all observers on earth to see the sun, moon, and the rest of the stars rising H15 and setting at the same times; rather, those living in the east invariably see them rise and set earlier, but those in the west later. For we find that the appearances of eclipses, and especially those of the moon, that occur at the same time are not recorded by all observers at the same hours; that is to say, at hours at equal intervals from noon. Instead, the hours recorded by the more easterly observers are always later than those by the more westerly observers. And since the difference in the hours is found to be proportional to the distances of the regions, one might reasonably suppose that the surface of the earth is spherical; it is the uniformity of its convexity, taken as a whole, that always produces occultations in proportion to successive distances. Were its shape different, this would not result, as one might see from the following considerations as well.

If it were concave, the stars would appear to rise earlier for more westerly observers. If it were planar, they would rise and set for all observers on the earth at the same times. And if it were a triangle or a square or any other polygonal figure, they would rise and set again in the same way and at the same time for those living on the same straight line. But in fact this is in no way seen to occur. And the following considerations make it clear that the earth could not even be cylindrical. For let its rounded surface be turned east H16 and west and the sides of its planar bases towards the poles of the universe (some might actually consider this possibility rather persuasive). For if it were cylindrical, none of the stars would always be visible for anyone living on its convex surface. Rather, all the stars would either rise and set for all observers, or the same stars, namely those at equal distances from each pole, would always be invisible for all observers. But in reality, the farther north we travel, the greater number of southerly stars are hidden and the greater number of northerly stars revealed. Hence it is clear that the convexity of the earth for this reason as well, by casting successive parts into shadow, fully proves that the shape is spherical. In addition, if we sail towards mountains or any elevated regions, from any direction and in any direction whatsoever, their magnitudes are observed to increase gradually, just as if they were rising up out of and had previously sunk into the sea itself, due to the convexity of the surface of the earth.

5. That the Earth Is in the Middle of the Heavens.

Now that this has been investigated, if someone should next treat of the position of the earth, he would realize that the appearances connected with it will only result if we should suppose that it is in the middle of the heavens, like a center of a sphere. For if this were, in fact, not the case, the earth ought either to be off the axis yet equally removed from each pole, or to be on the axis but to have receded towards one of the poles, or to be neither on the axis nor equally removed from each pole. H17

The following considerations, then, are in conflict with the first position. If the earth should be conceived of being displaced either above or below the axis relative to some observers, it would turn out for them that in right sphere[1] there would never be an equinox. For the part of the heavens above the earth and the part below the earth would always be divided unequally. And in oblique sphere,[2] there would again be either no equinox at all or it would not occur in the passage midway between the summer tropic and the winter tropic. For these distances would necessarily be unequal, because the equatorial circle, which is the greatest of the circles described by the poles of revolution of the heavens, would no longer be bisected by the horizon; instead, it would be one of the circles parallel to it and either more northerly or southerly. But it is agreed to by absolutely everyone that these distances really are invariably equal; for the augmentations, relative to the equinox, found in the longest day at the summer tropics are equal to the diminutions found in the shortest days at the winter tropics. And if its displacement should again be supposed to the east or west relative to some observers, for H18

---

1 *right sphere*: The technical term Ptolemy uses to signify that aspect of the celestial sphere that is seen by an observer located on the earth's equator.

2 *oblique sphere*: If the observer is located in some latitude north or south of the equator other than at the poles, the North-South axis of the celestial sphere will be inclined to the plane of his horizon. The sphere will be said to be oblique, and the observer is said to be "in oblique sphere." Ptolemy's term may also be rendered as "inclined sphere."

them it would result that neither would the magnitudes and distances of the stars appear equal at the eastern horizon and the western horizon, nor would the time from rising to culmination be equal to the time from culmination to setting. These results manifestly contradict the appearances in every way.

One might object again to the second position, in which the earth, on the axis, will be conceived to be displaced toward one of the poles. If this were the case, the plane of the horizon at every latitude would always make the part of the heavens above the earth and the part below the earth unequal; and it would make them unequal in different ways, both in themselves and relative to one another, at one displacement after another. Only in right sphere is the horizon able to bisect the sphere. But at an inclination which makes the nearer pole always visible, the horizon always diminishes the part above the earth, while it increases the part below the earth. Hence it results that even the great circle through the center of the signs of the zodiac[1] is divided H19 unequally by the plane of the horizon: a thing that is in no way observed to hold. Six of the zodiac signs always appear above the earth to all observers, while the remaining six are invisible. In turn, all of the latter again appear above the earth at the same time, while the rest do not appear along with them. Hence it is clear that the segments of the zodiac are also bisected by the horizon, because the same whole hemispheres are cut off above the earth at one time, but are cut off below the earth at another time.

And in general, if the earth did not hold its position right on the equatorial circle but displaced either to the north or south toward one of the poles, it would result that the rising shadows on gnomons at equinoxes would no longer even sensibly be in a straight line with the setting shadows on planes parallel to the horizon. But, in fact, this is plainly observed invariably to hold good. And it is immediately clear that the third position cannot be viable at all, since each contradiction in the first two positions will result for it.

To put it briefly, all of the regularity that is observed in the augmentations and diminutions of days and nights would be utterly confounded should the earth not be assumed in the middle. In addition, eclipses of the moon would not occur in all parts of the heavens at a position diametrically opposite the H20 sun, since the earth frequently would not occult them in their diametrically opposing passages, but rather at distances less than a hemisphere.

---

1 *the great circle through the center of the signs of the zodiac*: This great circle is the projection of the annual path of the sun onto the celestial sphere. The zodiac is an imaginary belt through the heavens contained by two circles, each parallel to this great circle and 8 degrees distant from it. The zodiac is divided into twelve houses or signs, each extending through 30 degrees of longitude. The sun, moon and five planets, as seen from the earth, are always within the zodiac.

6. That the Earth Has the Ratio of a Point to the Heavens.

There is in truth weighty proof that the earth has, to sensation, the ratio of a point compared with its distance to the sphere of the so-called fixed stars; from all parts of the earth the magnitudes and distances of the stars invariably appear the same and similar at the same times, just as observations of the same stars from different latitudes are not found discrepant in the least. What is more, the following has to be admitted. Gnomons placed on any part of the earth whatsoever and, furthermore, centers of armillary spheres[1] are equivalent to the true center of the earth. And they preserve the sightings and the revolutions of the shadows in such conformity with our hypotheses of the appearances that it is as if they really were occurring through the midpoint itself of the earth.

And a clear sign that these things are so is that the planes produced through our eyes, which we call horizons, invariably bisect the whole sphere of the heavens. But this would in fact not happen if the magnitude of the earth were perceptible in comparison with its distance to the heavens; only the plane produced through the point at the center of the earth would be able to bisect the sphere, while those produced through any chance surface of the earth would always make the portions beneath the earth greater than those above the earth. H21

7. That the Earth Does Not Move at All.

It will be shown by the same considerations as in the preceding that the earth does not make any motion at all in the above-mentioned oblique directions {H17.3 ff.}, or in general ever move from its place at the center. For the same things would result that result if it really were to have its position other than at the center. Therefore it seems to me that one would investigate to no purpose the causes for the movement to the center, since it is at once so clear from the appearances themselves that the earth occupies the middle place in the universe and that all heavy things move toward it. And the following consideration taken alone would be most conducive to such a conception. The earth, as we said, has been proven to be spherical and in the middle of the heavens. On every single part of it, the tiltings and movements of bodies with weight (I mean the tiltings and movements proper to them), always and invariably occur at right angles to the uninclined plane extended through the tangent point at incidence. Because this is so, it is also clear that if H22

[1] *armillary sphere*: An astronomical instrument composed of rings showing major circles of the celestial sphere. See Preliminaries (p. 8, above) for an illustration.

they were not resisted by the surface of the earth, they would surely arrive at the center itself. For the straight line that leads to the center is always at right angles to the plane through the section at the point of contact and tangent to the sphere.

All those who think it incredible that with such great weight the earth neither moves nor is carried anywhere seem to me to err in making the comparison by attending to their own experiences and not to what is unique of the universe. For such a thing would, I believe, no longer seem marvelous to them if they should observe that this magnitude of the earth, when compared with the entire body that contains it, has the ratio of a point to it. For in this light it will seem possible that what is proportionately the smallest is controlled and resisted, equally and at the same angle from every direction, by what is absolutely the greatest and possesses uniform elements. H23 For there is no down or up in the universe relative to it,[1] just as in a sphere no one would conceive this sort of thing at all. And, as far as the compounds in the universe are considered in terms of their proper and natural movement, light compounds with fine elements are fanned to the outside and to the circumference. But they seem to be rushing up for each of us, because what is above all our heads and is itself called "up" tends toward the containing surface. But heavy compounds with coarse elements are carried toward the middle and the center, yet they seem to fall down, because again what is at the feet of all of us and is called "down" itself tends toward the center of the earth. And it is with good reason that they collapse about the middle due to the resistance and pressure that are mutually equal and the same from every direction. Therefore the whole solid body of the earth is also with good reason conceived to be very great compared with what is carried towards it, insofar as the earth remains unaffected by the impetus of what are the very lightest things and welcomes, as it were, what falls upon it. But if it did have some motion in common, one and the same with the other heavy things, it would clearly outstrip all of them in being carried downward because its magnitude is so exceedingly great. And animals and things with a modicum H24 of weight would be left behind floating in the air, while the earth itself would promptly have fallen entirely out of the heavens themselves. But things of this sort, even only imagined, would appear utterly ridiculous.

Now some, with nothing more persuasive, as they suppose, to reply give their assent to these conclusions. But they think they would in no way contradict them if they should assume that, for argument's sake, the heavens are unmoved while the earth revolves around the same axis from west to east,

---

[1] H23.1 reads αὐτήν, which grammatically should go with γῆν, "earth" {H22.13}.

one revolution each day, most nearly. Or else they would even set both the earth and the heavens in motion to some extent, but only about the same axis, as we said, and proportionately to how they overtake one another.

They have failed to notice that, while perhaps nothing in their simpler conception[1] would, so far as the appearances of the stars go, hinder this from being so, such a thing would seem quite ridiculous based on what occurs around us ourselves and in the air. For let us concede to them a thing so contrary to nature. Let things that possess the finest elements and are the lightest either not move at all or in a manner indistinguishable from things of the opposite nature; although things in the air and with less fine elements manifestly move more quickly than all things that are more earthy. And let things that have the coarsest elements and are heaviest make a proper motion that is so swift and uniform; although, again, earthy things by agreement sometimes are not fit to be moved even by other things. But then they would agree that the revolution of the earth is the most violent of all the motions without exception around it, as it would make such a revolution in a brief time. So that everything not standing on the earth would appear always to make a single motion opposite to the earth's. And neither would a cloud nor anything that flies or is propelled appear passing to the east. For the earth would always outstrip everything and would be in advance in making its eastward motion, with the result that everything else would seem to recede to the west and rearwards. **H25**

For if they should say that the air also is revolved together with it in the same direction and the same speed, nonetheless the compounds[2] in it would always seem to lag behind the motion of the two. Or, if in fact, they too were revolved along with the air, as if united with it, they would no longer appear to advance or retreat; but they would always remain stationary, and would not deviate or move at all in their flights or trajectories. But in fact we so clearly see all of these things occurring without any slowness or swiftness at all attaching to them from the earth's not being stationary. **H26**

8. That There Are Two Different Kinds of Primary Motions in the Heavens.

Now it will suffice to have sketched out to this extent in summary fashion these hypotheses which are necessarily taken as preliminaries to the particular teachings as well as those that follow from them. They will be fully confirmed and corroborated by the actual agreement between the appearances and what

---

[1] τὴν ἁπλουστέραν ἐπιβολήν (H24.16). Ptolemy's phrase may here carry a dismissive connotation such as "simple-minded" or "simplistic".

[2] *compounds*: Theon here understands this term as referring to comets, shooting stars, and clouds (II.433.22–23).

we will next prove methodically. In addition to these things, one might think it right also to premise first this general consideration: that there are two different kinds of primary motions in the heavens. One is the motion by which everything is carried from east to west, always in the same way and at the same speed.[1] It produces its revolution on circles parallel to one another ⟨CD, QR⟩[2]; the circles are clearly described by the poles ⟨P_N, P_S⟩ of the sphere that revolves everything uniformly. The greatest circle among these ⟨QR⟩ is called the equatorial circle, because it alone is always bisected by a great circle, the horizon, and because the sun's revolution, when it coincides with it, everywhere sensibly produces an equinox. The H27 second kind is that in which the spheres of the stars, in the direction opposite to the previously stated movement, make certain motions around poles ⟨M, O⟩ that are different and not the same as those of the primary revolution.[3] And we assume that these things are so because, on the one hand, on the basis of observation on any single day, all things without exception in the heavens are seen rising, culminating, and setting in regions that are sensibly uniform and parallel to the equatorial circle—this is unique to the primary movement; and, on the other hand, on the basis of connected and more continuous observation, all the other stars[4] appear to maintain the distances between one another and, for the most part, the unique properties of the regions that are specific to the primary movement; but the sun, moon, and planets make certain movements

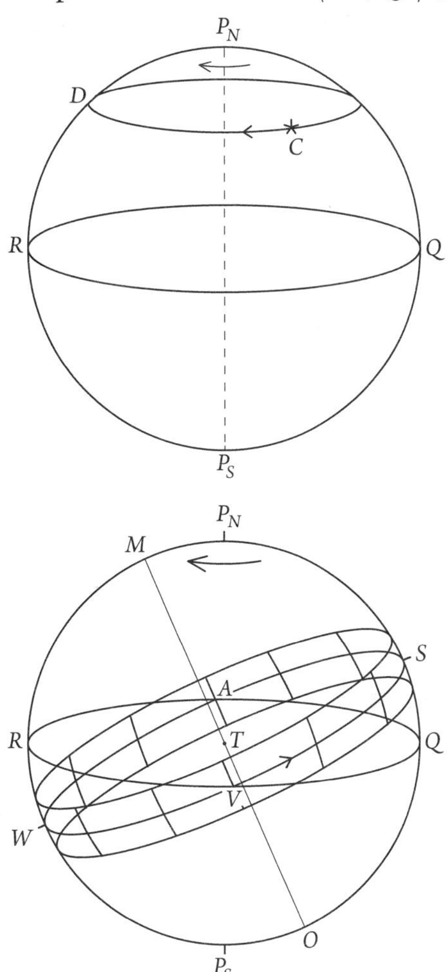

---

[1] Referring to the diurnal (daily) westward revolution of all of the heavenly bodies.

[2] The diagrams to which these and other bracketed letters refer have been supplied for this edition.

[3] *certain motions:* referring to the eastward motion of the sun, moon and planets through the signs of the zodiac.

[4] *all the other stars*: that is, the fixed stars.

that are complex and unequal to one another but all with the general motion to the east and rearwards of the stars that maintain their distances from one another and are revolved, as it were, by a single sphere.

If, then, this sort of motion of the planets were to occur on circles parallel to the equatorial circle, that is to say around the poles that produce the primary revolution, it would be sufficient to think that one and the same motion of them all is a consequence of the primary motion. For it would thus appear persuasive that the motions they have result from different time-lags and not from a contrary motion. But in reality, together with their motions to the east, they always appear displaced to the north and the south, and the magnitude of this displacement is not observed to be uniform. So that it might seem that this property belongs to them because of certain shocks.[1] But this motion is anomalous in light of the supposition of one primary revolution, but is produced in a regular manner by a circle ⟨*VSAW*⟩ oblique to the equatorial circle. Hence, this sort of circle that is one and same and specific to the planets is ascertained to be defined exactly and, as it were, described by the motion of the sun. And it is also traversed by the moon and the planets; they always revolve around it and do not randomly deviate from the displacement each of them cuts off from it in both directions. Now this is observed to be a great circle because the sun is more northerly and more southerly than the equatorial circle by an equal amount, and the eastward motions of all the planets occur, as we said, about one and same circle.[2] Hence it was necessary to posit this second different kind of general motion, which is produced around the poles ⟨*M*, *O*⟩ of the circle ascertained as oblique and in the direction ⟨*V*, *S*, *A*, *W*⟩ opposite to that of the primary movement.

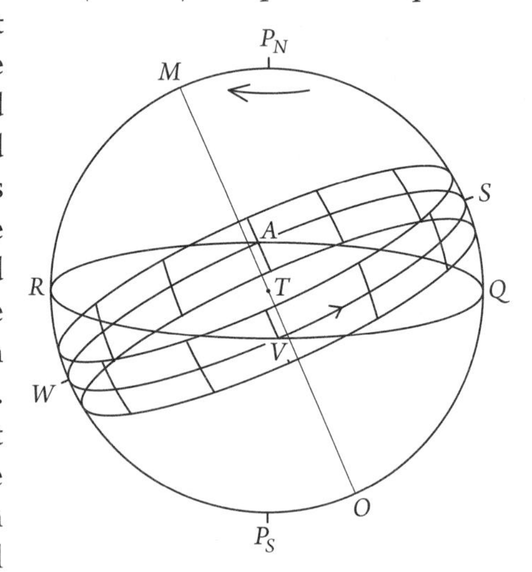

H28

H29

Now, if we conceive of a great circle ⟨$P_N MRWP_S OQS$⟩ through both pairs of poles of the previously mentioned circles which necessarily bisects each of them at right angles, that is to say the equatorial circle and the circle inclined to it, the oblique circle will have four points ⟨*V*, *S*, *A*, *W*⟩. Two of these ⟨*V*, *A*⟩ are produced by the equatorial circle diametrically opposite one another and are called equinoxes. Of these, the one ⟨*V*⟩ with its passage from

[1] *because of certain shocks*: that is, not naturally, but by force.

[2] *one and the same circle*: specifically, the mid-zodiac circle or *ecliptic*.

south to north is called the vernal equinox, while the opposite point ⟨A⟩ is called the autumnal equinox. And two points ⟨S, W⟩ are produced by the circle ⟨$P_N MRWP_S OQS$⟩ through both pairs of poles and are themselves also clearly diametrically opposite one another; and they are called tropics. Of these, the one ⟨W⟩ south of the equatorial circle is called the winter tropic, but the one ⟨S⟩ north the summer tropic.

And the one, primary movement that encompasses all the other movements will be conceived of as being circumscribed and, as it were, defined by the great circle ⟨$P_N MRWP_S OQS$⟩ that is described through both pairs of poles.[1] This circle is both revolved and revolves along with it all the other stars from east to west around the poles ⟨P_N, P_S⟩ of the equatorial circle that stand, as it were, upon the so-called meridian circle.[2] This circle differs from the previously mentioned one only in that it is not always described by the poles of the oblique circle and, further, because it is conceived of as being always at right angles to the horizon. And it is called meridian because this sort of position, which bisects both the hemisphere above the earth and the hemisphere below the earth, also contains the mean times of days and nights.[3] The second, manifold kind of motion is encompassed by the primary motion, and encompasses the spheres of all the planets. It is carried by the previously mentioned movement, as we said, and is counter-revolved in the opposite direction ⟨V, S, A, W⟩ around the poles ⟨M, O⟩ of the oblique circle. These poles themselves always stand on the circle ⟨$P_N MRWP_S OQS$⟩ that produces the primary motion; that is to say, the circle through both pairs of poles. And they are duly revolved with it; and in their motion in the direction opposite to the primary movement, they always preserve the same position[4] of the great, oblique circle ⟨$VSAW$⟩, described by their motion, to the equatorial circle ⟨$ARVQ$⟩.

H30

9. On the Particular Conceptions.

The general preliminary explanation in summary fashion would, then, contain the preceding sort of exposition of things that ought to be posited first. And as we are about to begin the particular demonstrations (among which we deem that demonstration to be primary through which it is

H31

1 *both pairs of poles*: specifically, the poles of the mid-zodiac circle and of the celestial equator.

2 *meridian circle*: Literally, the "mid-day" circle. The position of the meridian is fixed for any given place, and all of the heavenly bodies cross it twice daily—once above and once below the horizon. For more on the meridian and meridian circle, see Preliminaries (pp. 2–5 above.)

3 *the mean times of days and nights*: that is, noon and midnight.

4 *the same position*: that is, the same angle of inclination of the mid-zodiac circle to the equator.

ascertained what the magnitude really is of the arc between the previously mentioned poles ⟨P_N and M, or P_S and O⟩ on the circle ⟨$P_N M P_S O$⟩ that they describe), we see that it is necessary first to set out our treatment of the magnitude of chords in a circle in order to demonstrate everything geometrically once and for all.

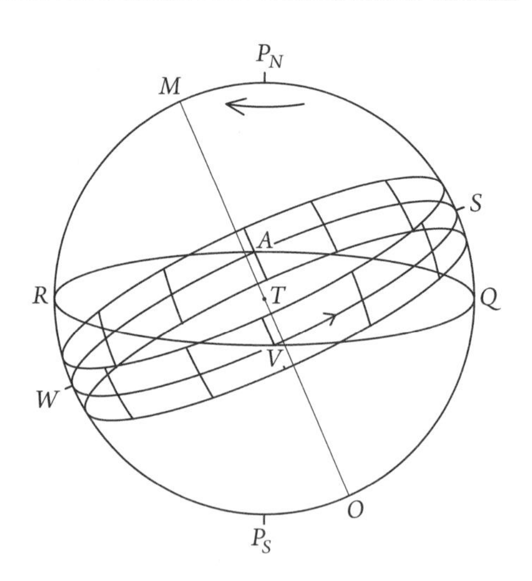

10. On the Magnitude of Chords in a Circle.

We will accordingly next produce a tabular exposition of the magnitudes of the chords for ready use. Having divided the circumference into 360 degrees, we will set alongside ⟨them⟩ the chords subtended by half-degree increases of arcs; that is to say, how many parts the chords are, with the diameter being divided into 120 parts. (The numerical utility of this will appear from the calculations themselves.) But first we will show how through the same few theorems we might make as systematic and rapid as possible the determination of the magnitudes of the chords, so that we not only have these magnitudes set out without fuss but also that we might easily submit them to scrutiny due to their systematic, geometric construction. In general we will use the numerical methods of the sexagesimal system[1] because of the inconvenience of fractions. Furthermore, we will follow out multiplications and divisions always aiming at what is approximate, and to the extent that the quantity neglected yields no significant difference from what is sensibly accurate. H32

Proposition 1.1

Now first, let semicircle ABC be on diameter ADC about center D, and let DB be drawn from D at right angles to AC, and let DC be bisected at E, and let EB be joined, and let EF be set equal to EB, and let FB be joined. I say that FD is a side of a decagon and BF a side of a pentagon.

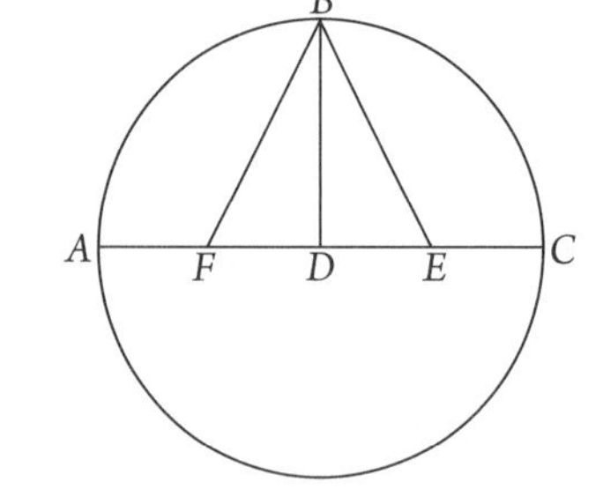

[1] The sexagesimal system is a base-60 number system, familiar to us from the division of hours into 60 minutes and minutes into sixty seconds. For further discussion of the sexagesimal system, see Preliminaries (pp. 9–11, above).

For, since straight line *DC* has been bisected at *E*, and there is added to
H33 it a certain straight line *DF*, the rectangle contained by *FC* and *FD* plus the square on *ED* is equal to the square on *EF* [Euc. 2.6];
that is to say, to the square on *BE*, since *BE* is equal to *FE*.

But the squares on *ED* and *DB* are equal to the square on *EB* [Euc. 1.47]. Therefore, the rectangle contained by *FC* and *FD* plus the square on *DE* is equal to the squares on *ED* and *BD*.

And when the square on *ED* is subtracted from both, the rectangle contained by *CF* and *FD* that remains is equal to the square on *DB*;
that is to say, to the square on *DC*.

Therefore, *FC* has been cut in extreme and mean ratio at *D* [Euc. 6.17, Euc. 6 def. 2].

Since, then, the side of the hexagon and the side of the decagon inscribed in the same circle are cut in extreme and mean ratio on the same straight line [Euc. 13.9];
and since the radius *CD* comprises the side of a hexagon [Euc. 4.15, por.],
therefore *DF* is equal to the side of the decagon.

Similarly, since the square on the side of the pentagon is equal to the square on side of the hexagon together with the square on the side of the decagon, when they are inscribed in the same circle [Euc. 13.10];
and since, in right-angled triangle *BDF*, the square on *BF* is equal to the
H34 square on *BD*, which is a side of a hexagon, together with the square on *DF* [Euc. 1.47], which is a side of a decagon,
therefore *BF* is equal to the side of the pentagon.

Calculation 1.1

Since, then, as I said {H31.16}, we are assuming the diameter of the circle to be 120 parts, through the preceding it arises that:
DE, being half of the radius, is 30 parts and the square on it is 900;
BD, the radius, is 60 parts, and the square on it is 3,600;
and added together the square on *EB*, that is to say the square on *EF*, is 4,500.

Therefore, *EF* in length will be 67;4,55 parts most nearly,
and the remainder *DF* 37;4,55 of the same parts.

Therefore, the side of the decagon, subtending an arc of 36 degrees where the circle is 360, will be 37;4,55 parts where the diameter is 120.

Calculation 1.2

Again, since *DF* is 37;4,55 parts, the square on it is 1,375;4,15, and the square on *DB* also is 3,600 of the same parts,
which, when added together, produce the square on *BF* as 4,975;4,15;
H35 therefore, *BF* in length will be 70;32,3 parts, most nearly.

And therefore the side of the pentagon, subtending 72 degrees, where the circle is 360, is 70;32,3 parts where the diameter is 120.

Calculation 1.3

And it is immediately clear that the side of the hexagon, subtending 60 degrees and being equal to the radius, is also 60 parts [Euc. 4.15, por.].

Calculations 1.4 – 1.5

Similarly, since the square on side of the square, subtending 90 degrees, is double the square on the radius [Euc. 1.47],
while the square on the side of the triangle, subtending 120 degrees,[1] is triple the square on the radius [Euc. 13.12],
and the square on the radius is 3,600 parts;
then the square on the side of the square will come out to be 7,200 parts and the square on the side of the triangle as 10,800 parts.

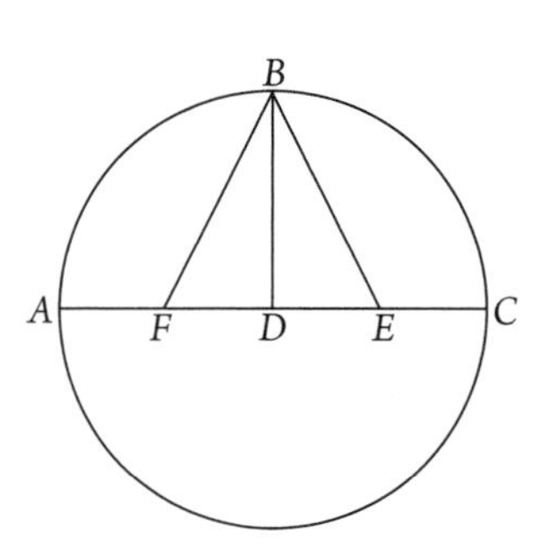

So that also the chord subtending 90 degrees will be 84;51,10 parts, most nearly, in length where the diameter is 120, while the chord subtending 120 degrees will be 103;55,23 of the same parts.

Proposition 1.2

Let these chords be thus obtained by us easily and by themselves, and it will be clear in consequence that the chords that subtend the supplementary arcs of the given chords in the semicircle are also readily given, because the squares on them added together make the square on the diameter [Euc. 1.47]. H36

Calculation 1.6

For example, since the chord subtending 36 degrees was shown to be 37;4,55 parts {H34.14}
and the square on this 1,375;4,15,
and the square on the diameter is 14,400 parts;
then the square on the chord that subtends the ⟨arc of⟩ 144 degrees supplementary to the semicircle will be the remaining 13,024;55,45 parts, while the chord itself will be 114;7,37 of the same parts, most nearly, in length. And similarly for the rest.

Calculations 1.7 – 1.10

We will next show how from these values the remaining particular chords will also be given, once we have set forth a little lemma that is quite useful for the present treatment.

Lemma 1.1

For let there be a circle with the chance inscribed quadrilateral, *ABCD*, and let *AC* and *DB* be joined. It must be proved that the rectangle contained by *AC* and *BD* is equal to the rectangle contained by *AB,DC* together with the rectangle contained by *AD,BC*.

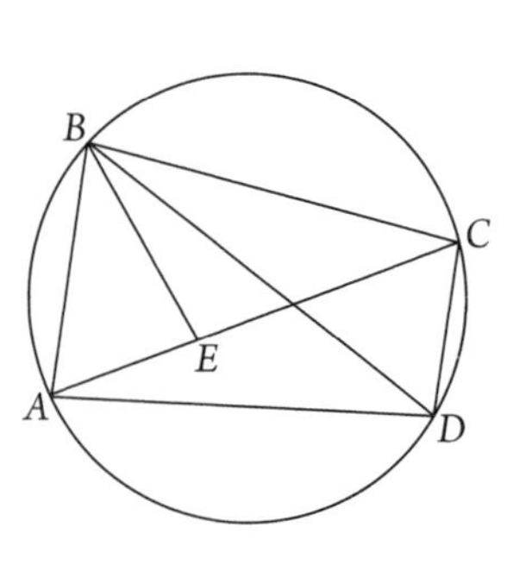

For let angle *ABE* be set equal to angle *DBC*.

Accordingly, if we add to both angle *EBD*, angle *ABD* will be equal to angle *EBC*.

Angle *BDA* is equal to angle *BCE*; for they subtend the same ⟨arc⟩ segment [Euc. 3.27]. H37

---

1 *the triangle subtending 120 degrees*: that is, the equilateral triangle.

Therefore, triangle *ABD* is equiangular with triangle *BCE*.

So that *BD* is proportionally to *DA* as *BC* is to *CE* [Euc. 6.4].

Therefore, the rectangle contained by *BC,AD* is equal to the rectangle contained by *BD,CE* [Euc. 6.16].

Again, since angle *ABE* is equal to angle *DBC*,
and angle *BAE* also is equal to angle *BDC*,
therefore triangle *ABE* is equiangular with triangle *BCD*;
therefore *BD* is proportionally to *DC* as *BA* is to *AE*.

Therefore the rectangle contained by *BA,DC* is equal to the rectangle contained by *BD,AE*.

It was also shown that the rectangle contained by *BC,AD* is equal to the rectangle contained by *BD,CE*.

And therefore the whole rectangle contained by *AC,BD* is equal to the rectangle contained by *AB,DC* together with the rectangle contained by *AD,BC* [Euc. 2.1]: which it was necessary to prove.

Proposition 1.3

H38

Once this has been set out first, let there be semicircle *ABCD* on diameter *AD*, and from *A* let two lines *AB* and *AC* be drawn through, and let each of them be given in magnitude in parts of which the diameter is given as 120, and let *BC* be joined.

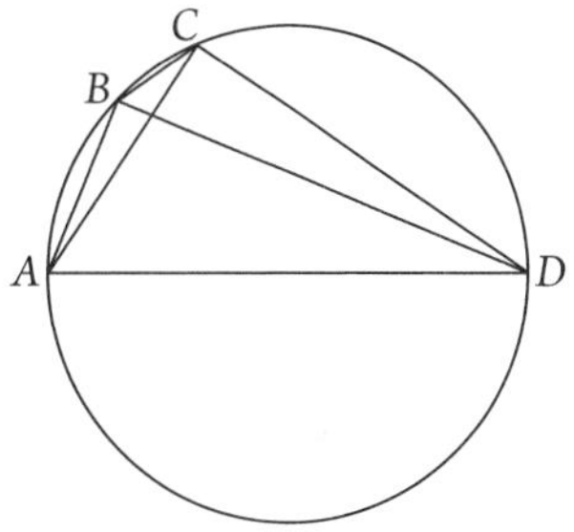

I say that *BC* too is given.

For let *BD* and *CD* be joined; therefore these also are evidently given, because they supplement the semicircle.

Since, then, the quadrilateral *ABCD* is in a circle, therefore the rectangle contained by *AB,DC* plus the rectangle contained by *AD,BC* is equal to the rectangle contained by *AC,BD*.

And both the rectangle contained by *AC,BD* and the rectangle contained by *AB,CD* are given;
and therefore the remaining rectangle contained by *AD,BC* is given.

And *AD* is a diameter;
therefore, chord *BC* also is given.

And it has become clear to us that, if two arcs and the chords subtending them are given, the chord that subtends the difference between the two arcs also will be given.

H39

Calculation 1.11

It is clear that, through this theorem, we will inscribe in the table many other chords based on the differences between the chords that have been given by themselves and especially that which subtends 12 degrees, since we have both the chord that subtends 60 degrees and the chord that subtends 72 degrees.

Proposition 1.4

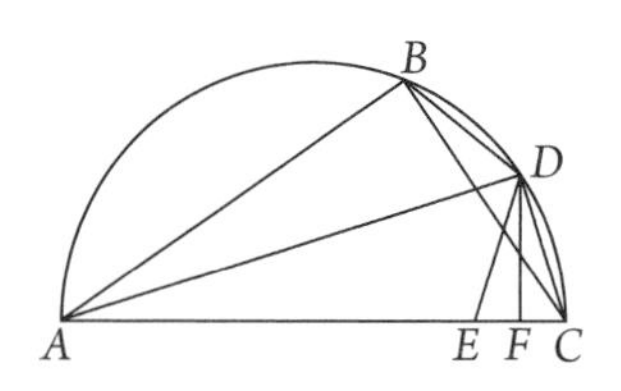

Again, let it be proposed, when a certain chord in a circle is given, to find the chord subtending half the subtended arc.

And let *ABC* be a semicircle on diameter *AC*, and *CB* be the given chord; and let arc *CB* be bisected at *D*, and let *AB*, *AD*, *BD*, and *DC* be joined, and from *D* let *DF* be drawn perpendicular to *AC*.

I say that *FC* is half the difference between chords *AB* and *AC*.

For let *AE* be set equal to *AB*, and let *DE* be joined.

Since *AB* is equal to *AE*, and *AD* is common, then the two lines *AB* and *AD* are equal to the two lines *AE* and *AD* respectively.

Angle *BAD* also is equal to angle *EAD* [Euc. 3.27];
and therefore base *BD* is equal to base *DE* [Euc. 1.4].

But *BD* is equal to *DC*; and therefore *DC* is equal to *DE*.

Accordingly, since in triangle *DEC*, which is isosceles, *DF* has been drawn from the vertex perpendicular to the base, *EF* is equal to *FC* [Euc. 1.26]. H40

But the whole *EC* is the difference between chords *AB* and *AC*;
therefore, *FC* is half the difference between the same chords.

So that, since, when the chord subtending arc *BC* is assumed, chord *AB* supplementary to the ⟨arc of the⟩ semicircle obviously is given as well,
FC, being half of the difference between *AC* and *AB*, also will be given.

But since, when *DF* is drawn perpendicular in right-angled triangle *ACD*, right-angled triangle *ACD* is equiangular with triangle *DCF*
and *CD* is to *CF* as *AC* is to *CD* [Euc. 6.4],
therefore the rectangle contained by *AC*,*CF* is equal to the square on *CD* [Euc. 6.16].

But the rectangle contained by *AC*,*CF* is given;
therefore, the square on *CD* also is given.

So that also the chord *CD*, subtending half the arc *BC*, will also be given in length.

Calculations 1.12 – 1.15

And again through this theorem a very great number of other chords will be gotten according to the halves of the ⟨arcs whose⟩ chords have been set out; and, above all, from the chord subtending 12 degrees, that subtending 6 degrees, that subtending 3 degrees, that subtending $1\frac{1}{2}$ degrees, and that subtending $\frac{3}{4}$ degree.

We find from our calculations that the chord that subtends $1\frac{1}{2}$ degrees is 1;34,15 parts, most nearly, where the diameter is 120, while that subtending $\frac{3}{4}$ degree is 0;47,8 of the same parts. H41

Proposition 1.5

Again, let circle *ABCD* be about diameter *AD* and its center *F*, and from *A* let there be intercepted two given arcs in succession, *AB* and *BC*, and let the chords that subtend them, *AB* and *BC*, be joined, themselves also given.

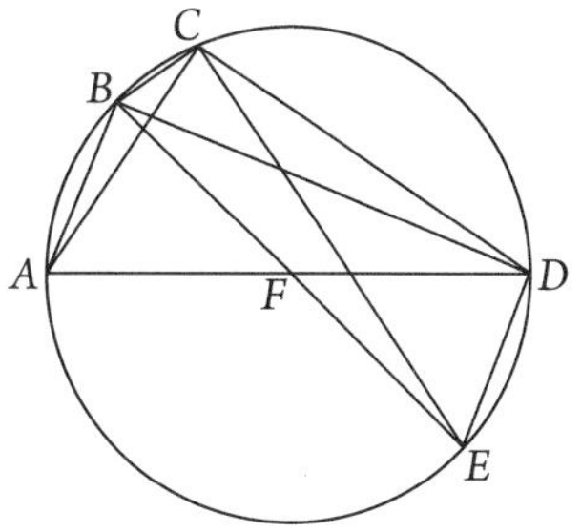

I say that, if we join *AC*, it itself will also be given.

For let *BFE* be drawn through *B* as a diameter of the circle, and let *BD*, *DC*, *CE*, and *DE* be joined.

Now it is immediately clear that *CE* also will be given through *BC*,
and both *BD* and *DE* will be given through *AB*.

And, because of the same considerations as in the preceding, since *BCDE* is a quadrilateral in a circle,
and *BD* and *CE* have been drawn through, the rectangle contained by the lines drawn through is equal to the sum of the rectangles contained by the opposite sides.

So that, since, the rectangle contained by *BD,CE* being given, the rectangle contained by *BC,DE* also is given,
therefore the rectangle contained by *BE,CD* is given.

H42

And diameter *BE* is given;
then *CD* that remains will be given,
and, for this reason, the chord supplementary to the ⟨arc of the⟩ semicircle, *CA*, will also be given.

So that, if two arcs and the chords that subtend them are given, the chord that subtends both arcs added together also will be given through this theorem.

Calculations 1.16 – 1.25

It is clear that by continuously adding the chord subtending $1\frac{1}{2}$ degrees to all the chords previously set out and calculating their sums, we will enter ⟨into the table⟩ simply all those chords that, by doubling, will be divisible by 3.

And only those arcs will still be left over that will fall, two in each case, between intervals of $1\frac{1}{2}$ degrees, since we enter the arc for each $\frac{1}{2}$ degree. So that, if we find the chord subtending $\frac{1}{2}$ degree, it will also, both by adding and subtracting the given chords that contain the intervals, completely fill up all the remaining intermediate chords for us. Since, when a certain chord has been given, as that subtending $1\frac{1}{2}$ degrees, that which subtends $\frac{1}{3}$ of the same arc is not given geometrically at all (if it really were possible, we would

H43

obviously also have the chord subtending $\frac{1}{2}$ degree). First, we will work out the chord subtending 1 degree on the basis of the chord subtending $1\frac{1}{2}$ degrees and the chord subtending $\frac{3}{4}$ degree, once we set down a little lemma. This little lemma, even if it is not, in general, able to determine quantities of chords, at least in the case of such exceedingly small quantities it might preserve a result indistinguishable from the quantities we have determined.

Lemma 1.2

For I say that, if in a circle two unequal chords are drawn through, the greater has to the lesser a lesser ratio than the arc on the greater chord has to the arc on the lesser chord.

For let there be circle *ABCD*, and let there be drawn through in it two unequal chords, *AB* the lesser, and *BC* the greater.

I say that chord *CB* has to chord *BA* a lesser ratio than arc *BC* has to arc *BA*.

For let angle *ABC* be bisected by *BD*, and let *AEC*, *AD*, and *CD* be joined.

And since angle *ABC* has been bisected by chord *BED*, chord *CD* is equal to chord *AD* [Euc. 3.26, Euc. 3.29], and *CE* is greater than *EA* [Euc. 6.3].

H44

Now let *DF* be drawn from *D* perpendicular to *AEC*.

Since, then, *AD* is greater than *ED*, while *ED* is greater than *DF* [Euc. 1.47], therefore the circle, described with *D* as center and with distance *DE*, will cut *AD* and will fall beyond *DF*.

Now let circle *GEH* be described and let *DFH* be extended.

And since sector *DEH* is greater than triangle *DEF*, while triangle *DEA* is greater than sector *DEG*, therefore triangle *DEF* has to triangle *DEA* a lesser ratio than sector *DEH* has to sector *DEG*.[1]

But straight line *EF* is to straight line *EA* as triangle *DEF* is to triangle *DEA* [Euc. 6.1], and angle *FDE* is to angle *EDA* as sector *DEH* is to sector *DEG*.[2]

Therefore, straight line *FE* has to straight line *EA* a lesser ratio than angle *FDE* has to angle *EDA*.

And therefore *componendo* straight line *FA* has to *EA* a lesser ratio than angle *FDA* has to angle *ADE*.

And also the doubles of the antecedents, straight line *CA* has to straight line *AE* a lesser ratio than angle *CDA* has to angle *EDA*.

H45

---

[1] *Lemma*: Given four magnitudes, $A > B$ and $C > D$, by a double application of Euc. 5.8 we can conclude that $A : D > A : C$ and $A : C > B : C$. If we extend Euc. 5.13 to the case where the first ratio is greater than the second instead of the same, we may conclude that $A : D > B : C$ and therefore $B : C < A : D$.

[2] The proof of this statement is not included in the original text of Euclid's *Elements,* but there is in many manuscripts an addendum to Euc. VI.33 stating that sectors of equal circles are (like the circumferences) in the same ratio as the angles at the center. Interestingly, in a note to VI.33 in his edition of the *Elements,* Heath writes, "These additions are clearly due to Theon, as may be gathered from his own statement in his commentary on the μαθηματικὴ σύνταξις [i. e., the *Almagest*] of Ptolemy, 'But that sectors in equal circles are to one another as the angles on which they stand, has been proved by me in my edition of the *Elements* at the end of the sixth book.'" Heath includes a proof in his comment on Euclid VI.33.

And *separando* straight line *CE* has to straight line *EA* a lesser ratio than angle *CDE* has to angle *EDA*.

But chord *CB* is to chord *BA* as straight line *CE* is to straight line *EA* [Euc. 6.3],
and arc *CB* is to arc *BA* as angle *CDB* is to angle *BDA* [Euc. 6.33];
therefore chord *CB* has to chord *BA* a lesser ratio than arc *CB* has to arc *BA*.

Calculation 1.26

Now, accordingly, with this lemma assumed, let there be circle *ABC*, and let there be drawn through in it two chords, *AB* and *AC*,
and let it be assumed, first, that chord *AB* subtends $\frac{3}{4}$ degree and *AC* 1 degree.

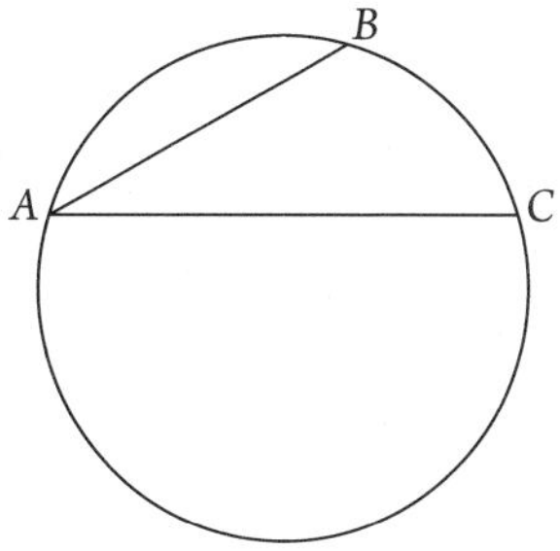

Since chord *AC* has to chord *AB* a lesser ratio than arc *AC* has to arc *AB*,
while arc *AC* is $1\frac{1}{3}$ arc *AB*,
therefore chord *AC* is less than $1\frac{1}{3}$ of *AB*.

But chord *AB* was shown to be 0;47,8 parts where the diameter is 120;
therefore chord *AC* is less than 1;2,50 of the same parts;
for 1;2,50 is, most nearly, $1\frac{1}{3}$ of 0;47,8.

H46

Again, in the same diagram, let chord *AB* be assumed as subtending 1 degree and chord *AC* $1\frac{1}{2}$ degrees.

Now, by the same considerations, since arc *AC* is $1\frac{1}{2}$ arc *AB*,
therefore chord *AC* is less than $1\frac{1}{2}$ of chord *BA*.

But we proved that chord *AC* is 1;34,15 parts where the diameter is 120;
therefore, chord *AB* is greater than 1;2,50 of the same parts;
for the preceding 1;34,15 parts are $1\frac{1}{2}$ of 1;2,50.

So that, since the chord subtending 1 degree has been shown to be both greater and less than the same parts, then we will evidently have this chord as 1;2,50 parts, most nearly, where the diameter is 120.[1]

Calculation 1.27

And because of what was previously shown, we will also have the chord subtending $\frac{1}{2}$ degree, which is found as 0;31,25, most nearly, of the same parts.

Calculations 1.28 – 1.29

And the remaining intervals, as we said, will be completely filled up. From the addition of the chord subtending $\frac{1}{2}$ degree with the chord subtending $1\frac{1}{2}$ degrees, in the first interval for instance, the chord subtending 2 degrees is demonstrated. And from the difference between the chord subtending

---

1 *both greater and less than the same number of parts*: It is only because the values found are approximations that the chord of 1° can be said to be "both greater and less than" 1;2,50^{p}. For $\frac{4}{3}$ of 0;47,8^{p} is 1;2,50,40^{p}; while $\frac{2}{3}$ of 1;34,15^{p} is 1;2,50,00^{p}. Thus the chord is actually shown to lie between 1;2,50,00^{p} and 1;2,50,40^{p}. The value midway between them, when rounded to the nearest second, is 1;2,50^{p}.

3 degrees and the chord subtending $\frac{1}{2}$ degrees, the chord subtending $2\frac{1}{2}$ degrees is also given; likewise for the rest as well.

The treatment, then, of chords in a circle would in this way, I suppose, be most easily handled. And in order, as I said {H31.20–32.1}, that we have the quantities of the chords readily set out for every need, we will attach tables by rows of 45, for common measure. Their first columns will contain the quantities of the arcs augmented by $\frac{1}{2}$ degree increments and their second columns the quantities of the chords set alongside the arcs, with the diameter assumed as 120 parts. Their third columns will contain $\frac{1}{30}$ of the augmentation of the chords for each $\frac{1}{2}$ degree increment, in order that, by having the average value of $\frac{1}{60}$ (which does not differ sensibly from the precise value), we are readily able to calculate as well the corresponding quantities for the fractional arcs that fall in between the $\frac{1}{2}$ degree intervals.[1] It is easy to understand that, through the above mentioned theorems, even if we are in uncertainty about a scribal error concerning one of the chords set out in rows in the table, we will readily perform the test and the correction; either from the chord subtending an arc double the arc being sought, or from the difference between some other given chords, or from the chord that subtends the supplementary arc to the semicircle. And the tabular diagram is as follows. H47

11. Table of Chords in a Circle. H48

arcs	chords			sixtieths			
½	0	31	25	0	1	2	50
1	1	2	50	0	1	2	50
1½	1	34	15	0	1	2	50
2	2	5	40	0	1	2	50
2½	2	37	4	0	1	2	48
3	3	8	28	0	1	2	48
3½	3	39	52	0	1	2	48
4	4	11	16	0	1	2	47
4½	4	42	40	0	1	2	47
5	5	14	4	0	1	2	46
5½	5	45	27	0	1	2	45
6	6	16	49	0	1	2	44
6½	6	48	11	0	1	2	43

[1] By taking thirtieths of the differences between entries in the table of chords, Ptolemy facilitates interpolation, or approximation, of the chord for every minute of arc. For example, the chord of 18;15° = 18;46,19^p (chord of 18°) + 15 times 0;1,2,2^p (thirtieth of difference), totalling 19;1,49,30^p.

arcs	chords			sixtieths			
7	7	19	33	0	1	2	42
7½	7	50	54	0	1	2	41
8	8	22	15	0	1	2	40
8½	8	53	35	0	1	2	39
9	9	24	51	0	1	2	38
9½	9	56	13	0	1	2	37
10	10	27	32	0	1	2	35
10½	10	58	49	0	1	2	33
11	11	30	5	0	1	2	32
11½	12	1	21	0	1	2	30
12	12	32	36	0	1	2	28
12½	13	3	50	0	1	2	27
13	13	35	4	0	1	2	25
13½	14	6	16	0	1	2	23
14	14	37	27	0	1	2	21
14½	15	8	38	0	1	2	19
15	15	39	47	0	1	2	17
15½	16	10	56	0	1	2	15
16	16	42	3	0	1	2	13
16½	17	13	9	0	1	2	10
17	17	44	14	0	1	2	7
17½	18	15	17	0	1	2	5
18	18	46	19	0	1	2	2
18½	19	17	21	0	1	2	0
19	19	48	21	0	1	1	57
19½	20	19	19	0	1	1	54
20	20	50	16	0	1	1	51
20½	21	21	11	0	1	1	48
21	21	52	6	0	1	1	45
21½	22	22	58	0	1	1	42
22	22	53	49	0	1	1	39
22½	23	24	39	0	1	1	36
23	23	55	27	0	1	1	33
23½	24	26	13	0	1	1	30
24	24	56	58	0	1	1	26
24½	25	27	41	0	1	1	22

arcs	chords			sixtieths			
25	25	58	22	0	1	1	19
25½	26	29	1	0	1	1	15
26	26	59	38	0	1	1	11
26½	27	30	14	0	1	1	8
27	28	0	48	0	1	1	4
27½	28	31	20	0	1	1	0
28	29	1	50	0	1	0	56
28½	29	32	18	0	1	0	52
29	30	2	44	0	1	0	48
29½	30	33	8	0	1	0	44
30	31	3	30	0	1	0	40
30½	31	33	50	0	1	0	35
31	32	4	8	0	1	0	31
31½	32	34	22	0	1	0	27
32	33	4	35	0	1	0	22
32½	33	34	46	0	1	0	17
33	34	4	55	0	1	0	12
33½	34	35	1	0	1	0	8
34	35	5	5	0	1	0	3
34½	35	35	6	0	0	59	57
35	36	5	5	0	0	59	52
35½	36	35	1	0	0	59	48
36	37	4	55	0	0	59	43
36½	37	34	47	0	0	59	38
37	38	4	36	0	0	59	32
37½	38	34	22	0	0	59	27
38	39	4	5	0	0	59	22
38½	39	33	46	0	0	59	16
39	40	3	25	0	0	59	11
39½	40	33	0	0	0	59	5
40	41	2	33	0	0	59	0
40½	41	32	3	0	0	58	54
41	42	1	30	0	0	58	48
41½	42	30	54	0	0	58	42
42	43	0	15	0	0	58	36
42½	43	29	33	0	0	58	31

arcs	chords			sixtieths			
43	43	58	49	0	0	58	25
43½	44	28	1	0	0	58	18
44	44	57	10	0	0	58	12
44½	45	26	16	0	0	58	6
45	45	55	19	0	0	58	0
45½	46	24	19	0	0	57	54
46	46	53	16	0	0	57	47
46½	47	22	9	0	0	57	41
47	47	51	0	0	0	57	34
47½	48	19	47	0	0	57	27
48	48	48	30	0	0	57	21
48½	49	17	11	0	0	57	14
49	49	45	48	0	0	57	7
49½	50	14	21	0	0	57	0
50	50	42	51	0	0	56	53
50½	51	11	18	0	0	56	46
51	51	39	42	0	0	56	39
51½	52	8	0	0	0	56	32
52	52	36	16	0	0	56	25
52½	53	4	29	0	0	56	18
53	53	32	38	0	0	56	10
53½	54	0	43	0	0	56	3
54	54	28	44	0	0	55	55
54½	54	56	42	0	0	55	48
55	55	24	36	0	0	55	40
55½	55	52	26	0	0	55	33
56	56	20	12	0	0	55	25
56½	56	47	54	0	0	55	17
57	57	15	33	0	0	55	9
57½	57	43	7	0	0	55	1
58	58	10	38	0	0	54	53
58½	58	38	5	0	0	54	45
59	59	5	27	0	0	54	37
59½	59	32	45	0	0	54	29
60	60	0	0	0	0	54	21
60½	60	27	11	0	0	54	12

arcs	chords			sixtieths			
61	60	54	17	0	0	54	4
61½	61	21	19	0	0	53	56
62	61	48	17	0	0	53	47
62½	62	15	10	0	0	53	39
63	62	42	0	0	0	53	30
63½	63	8	45	0	0	53	22
64	63	35	25	0	0	53	13
64½	64	2	2	0	0	53	4
65	64	28	34	0	0	52	55
65½	64	55	1	0	0	52	46
66	65	21	24	0	0	52	37
66½	65	47	43	0	0	52	28
67	66	13	57	0	0	52	19
67½	66	40	7	0	0	52	10
68	67	6	12	0	0	52	1
68½	67	32	12	0	0	51	52
69	67	58	8	0	0	51	43
69½	68	23	59	0	0	51	33
70	68	49	45	0	0	51	23
70½	69	15	27	0	0	51	14
71	69	41	4	0	0	51	4
71½	70	6	36	0	0	50	55
72	70	32	4	0	0	50	45
72½	70	57	26	0	0	50	35
73	71	22	44	0	0	50	26
73½	71	47	56	0	0	50	16
74	72	13	4	0	0	50	6
74½	72	38	7	0	0	49	56
75	73	3	5	0	0	49	46
75½	73	27	58	0	0	49	36
76	73	52	46	0	0	49	26
76½	74	17	29	0	0	49	16
77	74	42	7	0	0	49	6
77½	75	6	39	0	0	48	55
78	75	31	7	0	0	48	45
78½	75	55	29	0	0	48	34

arcs	chords			sixtieths			
79	76	19	46	0	0	48	24
79½	76	43	58	0	0	48	13
80	77	8	5	0	0	48	3
80½	77	32	6	0	0	47	52
81	77	56	2	0	0	47	41
81½	78	19	52	0	0	47	31
82	78	43	38	0	0	47	20
82½	79	7	18	0	0	47	9
83	79	30	52	0	0	46	58
83½	79	54	21	0	0	46	47
84	80	17	45	0	0	46	36
84½	80	41	3	0	0	46	25
85	81	4	15	0	0	46	14
85½	81	27	22	0	0	46	3
86	81	50	24	0	0	45	52
86½	82	13	19	0	0	45	40
87	82	36	9	0	0	45	29
87½	82	58	54	0	0	45	18
88	83	21	33	0	0	45	6
88½	83	41	4	0	0	44	55
89	84	6	32	0	0	44	43
89½	84	28	54	0	0	44	31
90	84	51	10	0	0	44	20
90½	85	13	20	0	0	44	8
91	85	35	24	0	0	43	57
91½	85	57	23	0	0	43	45
92	86	19	15	0	0	43	33
92½	86	41	2	0	0	43	21
93	87	2	42	0	0	43	9
93½	87	24	17	0	0	42	57
94	87	45	45	0	0	42	45
94½	88	7	7	0	0	42	33
95	88	28	24	0	0	42	21
95½	88	49	34	0	0	42	9
96	89	10	39	0	0	41	57
96½	89	31	37	0	0	41	45

arcs	chords			sixtieths			
97	89	52	27	0	0	41	33
97½	90	13	15	0	0	41	21
98	90	33	55	0	0	41	8
98½	90	54	29	0	0	40	55
99	91	14	56	0	0	40	42
99½	91	35	17	0	0	40	30
100	91	55	32	0	0	40	17
100½	92	15	40	0	0	40	4
101	92	35	42	0	0	39	52
101½	92	55	38	0	0	39	39
102	93	15	27	0	0	39	26
102½	93	35	11	0	0	39	13
103	93	54	47	0	0	39	0
103½	94	14	17	0	0	38	47
104	94	33	41	0	0	38	34
104½	94	52	58	0	0	38	21
105	95	12	9	0	0	38	8
105½	95	31	13	0	0	37	55
106	95	50	11	0	0	37	42
106½	96	9	2	0	0	37	29
107	96	27	47	0	0	37	16
107½	96	46	24	0	0	37	3
108	97	4	56	0	0	36	50
108½	97	23	20	0	0	36	36
109	97	41	38	0	0	36	23
109½	97	59	49	0	0	36	9
110	98	17	54	0	0	35	56
110½	98	35	52	0	0	35	42
111	98	53	43	0	0	35	29
111½	99	11	27	0	0	35	15
112	99	29	5	0	0	35	1
112½	99	46	35	0	0	34	48
113	100	3	59	0	0	34	34
113½	100	21	16	0	0	34	20
114	100	38	26	0	0	34	6
114½	100	55	28	0	0	33	52

arcs	chords			sixtieths			
115	101	12	25	0	0	33	39
115½	101	29	15	0	0	33	25
116	101	45	57	0	0	33	11
116½	102	2	33	0	0	32	57
117	102	19	1	0	0	32	43
117½	102	35	22	0	0	32	29
118	102	51	37	0	0	32	15
118½	103	7	41	0	0	32	0
119	103	23	44	0	0	31	46
119½	103	39	37	0	0	31	32
120	103	55	23	0	0	31	18
120½	104	11	2	0	0	31	4
121	104	26	34	0	0	30	49
121½	104	41	59	0	0	30	35
122	104	57	16	0	0	30	21
122½	105	12	26	0	0	30	7
123	105	27	30	0	0	29	52
123½	105	42	26	0	0	29	37
124	105	57	14	0	0	29	23
124½	106	11	55	0	0	29	8
125	106	26	29	0	0	28	54
125½	106	40	56	0	0	28	39
126	106	55	15	0	0	28	24
126½	107	9	27	0	0	28	10
127	107	23	32	0	0	27	56
127½	107	37	30	0	0	27	40
128	107	51	20	0	0	27	25
128½	108	5	2	0	0	27	10
129	108	18	37	0	0	26	56
129½	108	32	5	0	0	26	41
130	108	45	25	0	0	26	26
130½	108	58	38	0	0	26	11
131	109	11	44	0	0	25	56
131½	109	24	42	0	0	25	41
132	109	37	32	0	0	25	26
132½	109	50	15	0	0	25	11

arcs	chords			sixtieths			
133	110	2	50	0	0	24	56
133½	110	15	18	0	0	24	41
134	110	27	39	0	0	24	26
134½	110	39	52	0	0	24	10
135	110	51	57	0	0	23	55
135½	111	3	54	0	0	23	40
136	111	15	44	0	0	23	25
136½	111	27	26	0	0	23	9
137	111	39	1	0	0	22	54
137½	111	50	28	0	0	22	39
138	112	1	47	0	0	22	24
138½	112	12	59	0	0	22	8
139	112	24	3	0	0	21	53
139½	112	35	0	0	0	21	37
140	112	45	48	0	0	21	22
140½	112	56	29	0	0	21	7
141	113	7	2	0	0	20	51
141½	113	17	25	0	0	20	36
142	113	27	44	0	0	20	20
142½	113	37	54	0	0	20	4
143	113	47	26	0	0	19	49
143½	113	57	50	0	0	19	33
144	114	7	37	0	0	19	17
144½	114	17	15	0	0	19	2
145	114	26	46	0	0	18	46
145½	114	36	9	0	0	18	30
146	114	45	24	0	0	18	14
146½	114	54	31	0	0	17	59
147	115	3	30	0	0	17	43
147½	115	12	22	0	0	17	27
148	115	21	6	0	0	17	11
148½	115	29	41	0	0	16	55
149	115	38	9	0	0	16	40
149½	115	46	29	0	0	16	24
150	115	54	40	0	0	16	8
150½	116	2	44	0	0	15	52
151	116	10	40	0	0	15	36

arcs	chords			sixtieths			
151½	116	18	28	0	0	15	20
152	116	26	8	0	0	15	4
152½	116	33	40	0	0	14	48
153	116	41	4	0	0	14	32
153½	116	48	20	0	0	14	16
154	116	55	28	0	0	14	0
154½	117	2	28	0	0	13	44
155	117	9	20	0	0	13	28
155½	117	16	4	0	0	13	12
156	117	22	40	0	0	12	56
156½	117	29	8	0	0	12	40
157	117	35	28	0	0	12	24
157½	117	41	40	0	0	12	7
158	117	47	43	0	0	11	51
158½	117	53	39	0	0	11	35
159	117	59	27	0	0	11	19
159½	118	5	7	0	0	11	3
160	118	10	37	0	0	10	47
160½	118	16	1	0	0	10	31
161	118	21	16	0	0	10	14
161½	118	26	23	0	0	9	58
162	118	31	22	0	0	9	42
162½	118	36	13	0	0	9	25
163	118	40	55	0	0	9	9
163½	118	45	30	0	0	8	53
164	118	49	56	0	0	8	37
164½	118	54	15	0	0	8	20
165	118	58	25	0	0	8	4
165½	119	2	26	0	0	7	48
166	119	6	20	0	0	7	31
166½	119	10	6	0	0	7	15
167	119	13	44	0	0	6	59
167½	119	17	13	0	0	6	42
168	119	20	34	0	0	6	26
168½	119	23	47	0	0	6	10
169	119	26	52	0	0	5	53
169½	119	29	49	0	0	5	37

arcs	chords			sixtieths			
170	119	32	37	0	0	5	20
170½	119	35	17	0	0	5	4
171	119	37	49	0	0	4	48
171½	119	40	13	0	0	4	31
172	119	42	28	0	0	4	14
172½	119	44	35	0	0	3	58
173	119	46	35	0	0	3	42
173½	119	48	26	0	0	3	26
174	119	50	8	0	0	3	9
174½	119	51	43	0	0	2	53
175	119	53	10	0	0	2	36
175½	119	54	27	0	0	2	20
176	119	55	38	0	0	2	3
176½	119	56	39	0	0	1	47
177	119	57	32	0	0	1	30
177½	119	58	18	0	0	1	14
178	119	58	55	0	0	0	57
178½	119	59	24	0	0	0	41
179	119	59	44	0	0	0	25
179½	119	59	56	0	0	0	9
180	120	0	0	0	0	0	0

H63

12. On the Arc Between the Tropics.

H64.2

Now that the magnitude of chords in a circle has been set out, the first task would be, just as we said {H30.21–31.3}, to show how much the circle ⟨*VSAW*⟩, which is oblique and through the middle of the signs of the zodiac, is inclined towards the equatorial circle ⟨*VQAR*⟩; that is to say, what ratio the great circle ⟨MP_NOP_S⟩ through both pairs of the poles ⟨*M*, *O* and P_N, P_S⟩ we have set out has to the arc ⟨MP_N or OP_S⟩ intercepted on it between the poles. The point ⟨*Q* or *R*⟩ on the equatorial circle is also

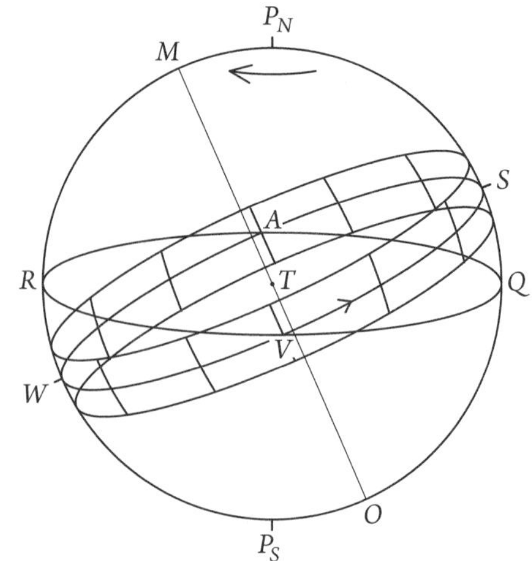

clearly equally distant from each of the tropic points ⟨S, W⟩ by an arc ⟨QS or RW⟩ equal to this arc. This is directly ascertained by mechanical means for us through the following simple sort of instrument.

For we will construct a ring of bronze, moderate in size, that has been turned accurately, and has the same width and breadth. We will employ this as a meridian circle[1] once we divide it into the 360 assumed parts of a great circle and each of these into as many fractions as is possible. Then we will attach a different, more delicate small ring within the one mentioned in such a way that their sides remain on a single surface, and that the smaller ring can be rotated within the larger ring without obstruction in the same plane northwards and southwards. We will then add at some two diametrically opposite segments of the smaller ring, on each side, short small prisms that incline equally towards each other and the center of the rings, once we have precisely set along the middle of their breadth delicate, small pointers that touch the side of the larger, divided ring. Now we will also securely fasten this, for specific needs, upon a moderate, small stanchion, once we set down in the open air the base of the small stanchion on a foundation that is not inclined to the plane of the horizon. And we will watch that the plane of the rings is perpendicular to the plane of the horizon, but parallel to the plane of the meridian circle. The first of these is handled through a plumb-line, which has been suspended from the point that will be at the vertex[2] and is observed until, from the correction of the bases, it makes its inclination towards its diametrically opposite point. But the second is handled through a meridian line[3] that is distinctly found in the plane beneath the small stanchion and by turning the rings obliquely

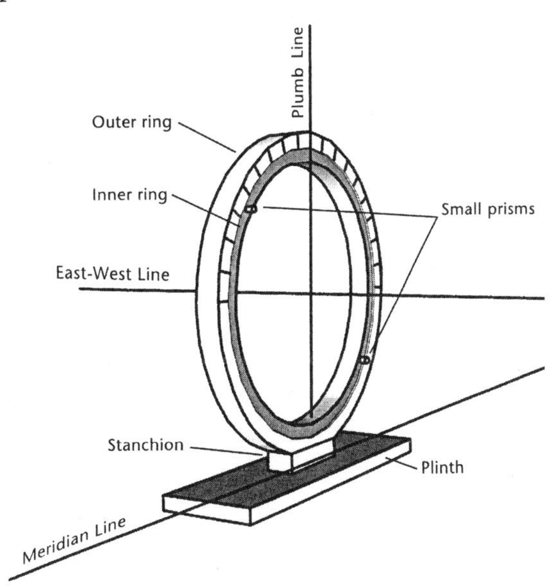

A representation of Ptolemy's "ring of bronze"

H65

---

[1] The meridian circle for a given place is a great circle of the celestial sphere passing through its poles and the zenith of that place.

[2] *the point that will be at the vertex*: that is, the point of the outer ring that corresponds to the zenith.

[3] Ptolemy does not explain how he determines the meridian line. For discussion of the meridian, see Preliminaries (pp. 2–3) and Epitome (pp. 18–19), above..

until their plane is sighted parallel to the line. Now when such a position is produced, we would look for the displacement of the sun northwards and southwards, while we turned the small, inner ring at mid-days until the whole, small prism below is cast into shadow by the whole prism above. And when this occurred, the extremities of the small pointers would clearly mark for us by how many parts the center of the sun, on the meridian circle, is separated on each occasion from the point at the vertex.[1] H66

We made this sort of observation in a still easier way by preparing, instead of the rings, a stone or wood plinth that was squared and undistorted, having, however, one of its sides uniform and precisely cut. Using some point upon it at one of the corners, we described a quadrant of a circle, joined from the point at the center up to the described arc the straight lines which contain the right angle subtending the quadrant, and similarly divided the arc into 90 degrees and their fractions. Next, we inserted upon one of the straight lines, which will be perpendicular to the plane of the horizon and have its position towards the south, two upright and completely equal small cylinders, uniformly lathed: one upon the point at the center precisely around the mid-point, the other at the bottom limit of the chord. Then we set this inscribed side of the plinth alongside the meridian line, that is drawn through in the underlying plane, so that it too has its position parallel to the plane of the meridian circle. And by means of a plumb-line through the small cylinders, we accurately fixed the straight line through them, true and perpendicular to the plane of the horizon. Again, when certain small bases[2] were corrected as required, we would observe in just the same way at mid-days the shadow that arose from the small cylinder at the center. By setting something alongside the inscribed arc with a view to its place appearing more clearly and by marking the middle of this shadow, we would obtain the segment on it, of the arc of the quadrant, that clearly indicated the latitudinal passage of the sun upon the meridian circle. H67

A modern example of Ptolemy's quadrant

Now, from these sorts of observations and, most especially, from observation of the tropics that we examined over greater periods, the

---

[1] *vertex*: high point; here referring to the zenith.

[2] *certain small bases*: that is, shims used for leveling the plinth.

indicator generally cut off from the point at the vertex[1] equal and identical segments of the meridian circle at both the summer and the winter tropics. Hence we obtained the arc from the most northerly limit to the most

H68

southerly, which is that between the tropic segments. This always proved to be 47 and more than $\frac{2}{3}$ but less than $\frac{3}{4}$ parts; through this is yielded roughly the same ratio as that of Eratosthenes, which Hipparchus used. For the arc between the tropics is 11 parts, most nearly, where the meridian circle is 83 {H77.24–25}.

From the preceding observation the latitudes of the zones of habitation, in which we make the observations, also immediately become easy to obtain. For we obtain both the point between the two limits, which is on the equatorial circle, and the arc between this and the point at the vertex; its poles too are clearly removed from the horizon by an arc equal to this arc.

13. Preliminaries to the Spherical Proofs.[2]

Since what follows is to demonstrate as well the particular quantities of the arcs, of great circles described through the poles of the equatorial circle, that are intercepted between the equatorial and the mid-zodiac circle, we will first set out short, useful little lemmas. Through these we will fashion the greatest number, more or less, of the proofs of spherical theorems in a way that is, to the greatest extent possible, quite simple and systematic.

Lemma 1.3

Now let two straight lines, *BE* and *CD*,

H69

produced to two straight lines *AB* and *AC*, cut each other at point *F*.

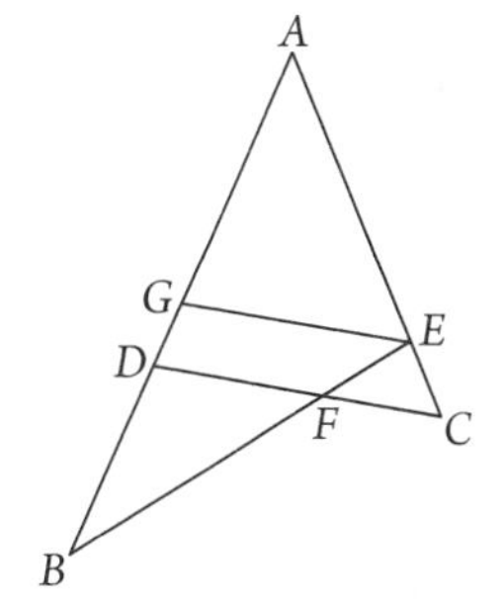

I say that the ratio of *CA* to *AE* is compounded from the ratio of *CD* to *DF* and the ratio of *FB* to *BE*.

For let *EG* be drawn through *E* parallel to *CD*.

Since *CD* and *EG* are parallel, the ratio of *CA* to *EA* is the same as the ratio of *CD* to *EG*.

And *FD* is an intermediary;

therefore the ratio of *CD* to *EG* will be compounded from the ratio of *CD* to *DF* and the ratio of *DF* to *GE*;

---

[1] *vertex*: high point; here, the point where the sun crosses the meridian.

[2] The following Lemmas 1.3 and 1.4 are generally known as the (planar) Menelaus Theorems, after the Greek geometer Menelaus of Alexandria (c.70–c.130). Although Menelaus did not originate the planar theorems, he used them to prove a corresponding spherical theorem, which appears in the present work as Proposition 1.6 below. For a convenient mnemonic formulation of the planar Menelaus Theorems, see Appendix 3.

so that also the ratio of CA to AE is compounded from the ratio of CD to DF and the ratio of DF to GE.

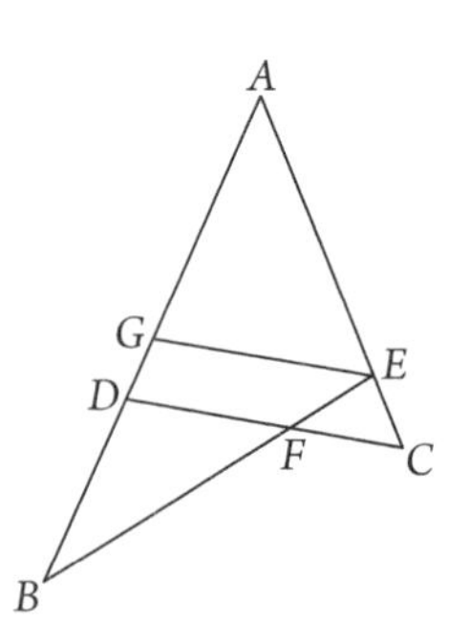

And the ratio of DF to GE is also the same as the ratio of FB to BE,
because EG and FD again are parallel;
therefore the ratio of CA to AE is compounded from the ratio of CD to DF and the ratio of FB to BE:
which it was proposed to prove.

Lemma 1.4

By the same considerations it will be proved that, *separando* also, the ratio of CE to EA is compounded from the ratio of CF to DF and the ratio of BD to BA, when a line is drawn through A parallel to EB, and CDG is extended to it.

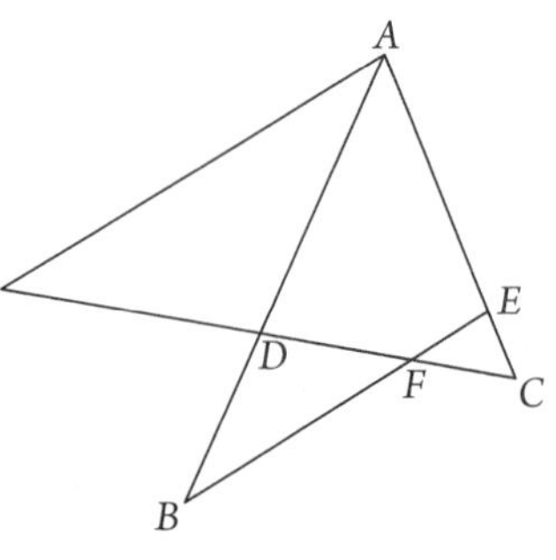

For since, again, AG is parallel to EF,
CF is to FG as CE is to EA.

H70

But when DF is taken as an intermediary,
the ratio of CF to FG is compounded from the ratio of CF to DF and the ratio of DF to FG.

And the ratio of DF to FG is the same as ratio of DB to BA,
because BA and FG have been produced to the parallels AG and FB.

Therefore, the ratio of CF to FG is compounded from the ratio of CF to DF and the ratio of DB to BA.

But the ratio of CE to EA is the same as the ratio of CF to FG;
and therefore the ratio of CE to EA is compounded from the ratio of CF to DF and the ratio of DB to BA:
which it was necessary to prove.

Lemma 1.5

Again let there be circle ABC, whose center is D, and let there be taken on its circumference three random points, A, B, and C, so that each of the arcs AB and BC is less than a semicircle; and let the like be understood for successively taken arcs.

And let AC and DEB be joined.

I say that straight line AE is to straight line EC as the chord subtending the arc double the arc AB is to the chord subtending the arc double the arc BC.

H71

For from points A and C let AF and CG be drawn perpendicular to DB.

Since *AF* is parallel to *CG*, and straight line *AEC* has been produced to them,
AE is to *EC* as *AF* is to *CG*.

But the ratio of *AF* to *CG* and the ratio of the chord that subtends the arc double the arc *AB* to the chord subtending the arc double the arc *BC* is the same; for each is half of the other.

And therefore the ratio of *AE* to *EC* is the same as the ratio of the chord that subtends the arc double the arc *AB* to the chord that subtends the arc double the arc *BC*:
which it was necessary to prove.

Lemma 1.6

It immediately follows that, even if the whole arc *AC* and the ratio of the chord that subtends the arc double the arc *AB* to the chord that subtends the arc double the arc *BC* are given, each of the arcs *AB* and *BC* will also be given.

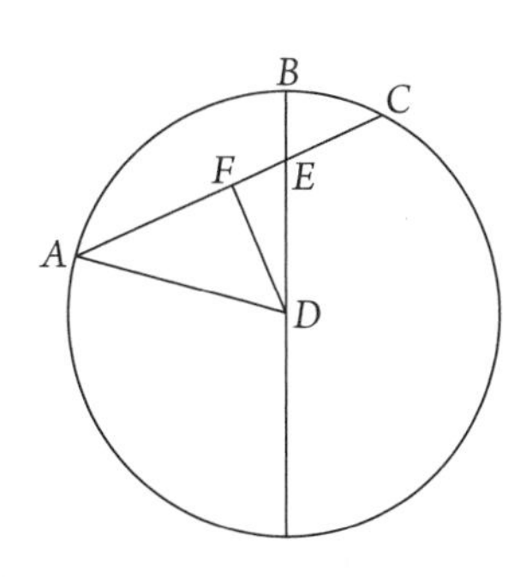

For when the same diagram is set out, let *AD* be joined, and let *DF* be drawn from *D* perpendicular to *AEC*.

H72

The fact, then, that, when arc *AC* is given, both angle *ADF*, which subtends half of it, and the whole triangle *ADF* will be given, is clear.

And since, when the whole chord *AC* is given, the ratio of *AE* to *EC* also is assumed as being the same as the ratio of the chord that subtends the arc double the arc *AB* to the chord subtending the arc double the arc *BC*, both *AE* and the remainder *FE* will be given.

And for this reason, as *DF* also is given, both angle *EDF* of right-angled triangle *EDF* and the whole angle *ADB* will be given;
so that also, arc *AB* as well as its remainder *BC* will also be given:
which it was necessary to prove.

Lemma 1.7

Again, let circle *ABC* be around center *D*, and upon its circumference let there be taken three points, *A*, *B*, and *C*, so that each of the arcs *AB* and *AC* is less than a semicircle; and let the like be understood for successively taken arcs. And let *DA* and *CB*, being joined, be extended and meet at point *E*.

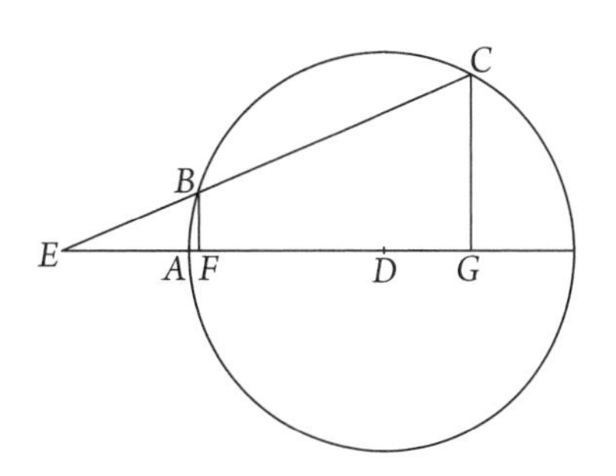

H73

I say that straight line *CE* is to straight line *BE* as the chord subtending the arc double the arc *CA* is to the chord subtending the arc double the arc *AB*.

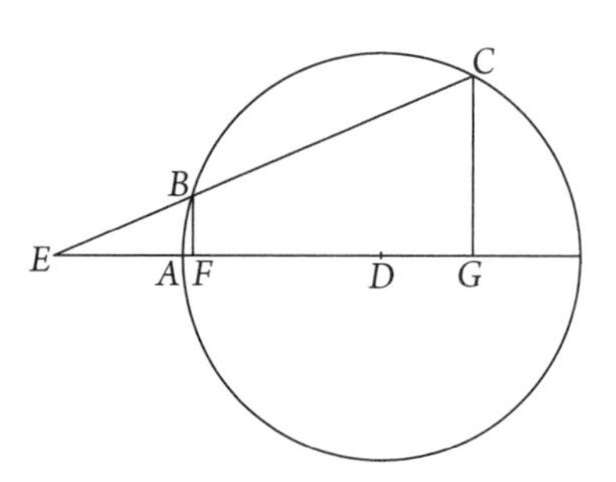

For, in like manner with the prior little lemma, if from points *B* and *C* we draw *BF* and *CG* perpendicular to *DA*, because they are parallel, *CE* will be to *EB* as *CG* is to *BF*;
so that also,
CE is to *EB* as the chord subtending the arc double the arc *CA* is to the chord subtending the arc double the arc *AB*:
which it was necessary to prove.

Lemma 1.8

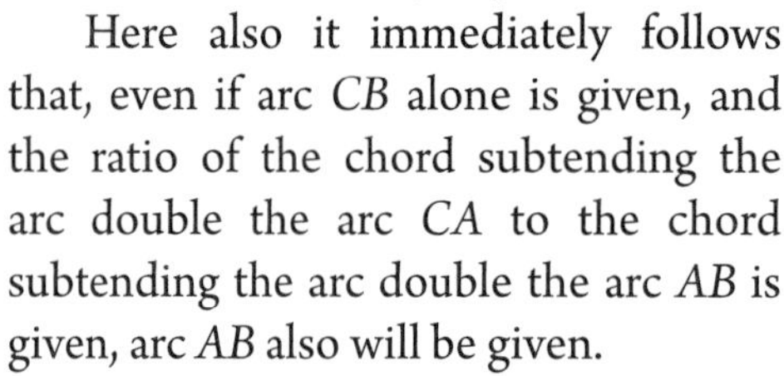

Here also it immediately follows that, even if arc *CB* alone is given, and the ratio of the chord subtending the arc double the arc *CA* to the chord subtending the arc double the arc *AB* is given, arc *AB* also will be given.

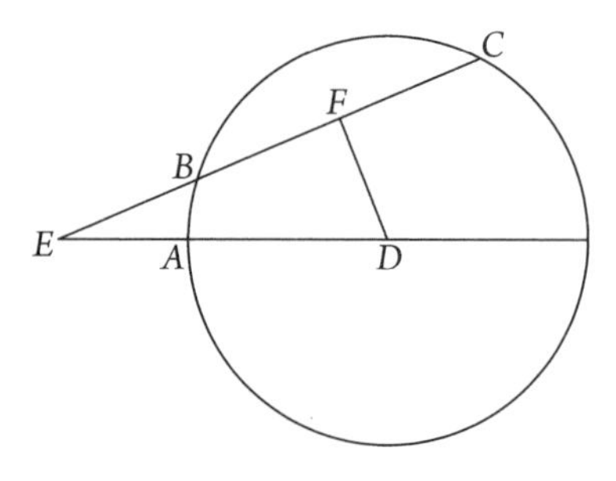

For, again, in the like diagram, when *DB* is joined and *DF* is drawn perpendicular to *BC*, angle *BDF*, subtending half the arc *BC*, will be given;

H74

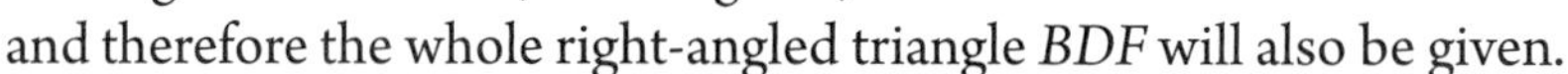

and therefore the whole right-angled triangle *BDF* will also be given.

And since the ratio of *CE* to *EB* also is given,[1] and, furthermore, the straight line *CB*,
then *EB* and, furthermore, the whole *EBF* will be given.

So that also, since *DF* is given,
both angle *EDF* of the same right-angled triangle and the remaining angle *EDB* will be given.

So that also arc *AB* will be given.

Proposition 1.6

When these lemmas are presupposed, let there be inscribed on a spherical surface arcs of great circles, so that two arcs *BE* and *CD*, drawn to two arcs *AB* and *AC*, cut each other at point *F*; and let each of them be less than a semicircle. Let the same thing also be understood for all the diagrams.

Now, I say that the ratio of the chord subtending the arc double the arc *CE* to the chord subtending the arc double the arc *EA* is compounded from the ratio of the chord subtending the arc double the arc *CF* to the chord subtending the arc double the arc *FD* and the ratio of the chord subtending

1 *the ratio of CE to EB also is given*: For the ratio of the chord subtending double the arc *CA* to the chord subtending double the arc *AB* was given; and it is equal to the ratio of *CE* to *EB* by Lemma 1.7.

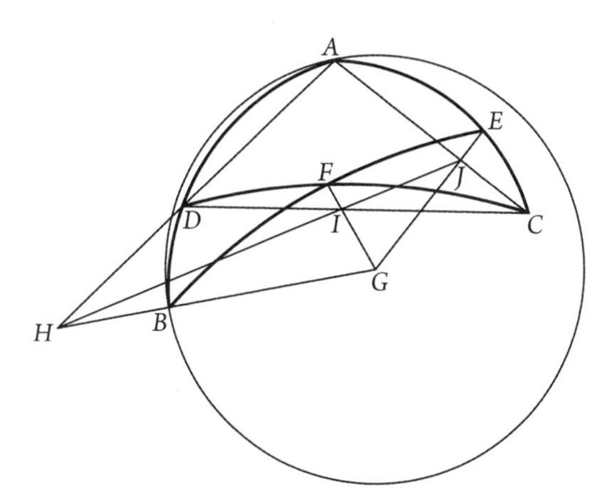

the arc double the arc *DB* to the chord subtending the arc double the arc *BA*.[1]

For let the center of the sphere be taken, and let it be *G*,
and let *GB*, *GF*, and *GE* be drawn from *G* to the ⟨inter⟩sections *B*, *F*, and *E* of the circles.

And let *AD*, being joined, be extended and let it meet *GB*, itself also extended, at point *H*,

H75 and similarly let *DC* and *AC*, being joined, cut *GF* and *GE* at point *I* and point *J*.

Now, points *H*, *I*, and *J* are upon one straight line, because they are simultaneously in two planes: that of the triangle *ACD* and that of circle *BFE*.

When joined, this line makes *HJ* and *CD*, that have been produced to two straight lines *HA* and *CA*, cut each other at point *I*.

Therefore, the ratio of *CJ* to *JA* is compounded from the ratio of *CI* to *ID* and the ratio of *DH* to *HA* {H69.21}.

But the chord subtending the arc double the arc *CE* is to the chord subtending the arc double the arc *EA* as *CJ* to *JA*.

And the chord subtending the arc double the arc *CF* is to the chord subtending the arc double the arc *FD* {H70.17} as *CI* is to *ID*.

And the chord subtending the arc double the arc *DB* is to the chord subtending the arc double the arc *BA* {H72.17} as *HD* is to *HA*.

And therefore the ratio of the chord subtending the arc double the arc *CE* to the chord subtending the arc double the arc *EA* is compounded from

H76 the ratio of the chord subtending the arc double the arc *CF* to the chord subtending the arc double the arc *FD* and the ratio of the chord subtending the arc double the arc *DB* to the chord subtending the arc double the arc *BA*.

---

1 This proposition is generally known as the (spherical) Menelaus Theorem. See Appendix 3.

Proposition 1.7

Now, by the same considerations and just as in the case of the plane diagram of straight lines[1] {H68.3}, it is proved that the ratio of the chord subtending the arc double the arc *CA* to the chord subtending the arc double the arc *EA* also is compounded from the ratio of the chord subtending the arc double the arc *CD* to the chord subtending the arc double the arc *DF* and the ratio of the chord subtending the arc double the arc *FB* to the chord subtending the arc double the arc *BE*: which it was proposed to prove.

14. On the Arcs Between the Equatorial Circle and the Oblique Circle.

Now, since this theorem has been set out first, we will first prove the arcs under consideration as follows.

Proposition 1.8
Calculation 1.30

For let *ABCD* be the circle through both pairs of poles, of the equatorial circle and the mid-zodiac circle, and *AEC* the semicircle of the equatorial circle, and *BED* the semicircle of the mid-zodiac circle, and point *E* their intersection at the vernal equinox, so that *B* is the winter and *D* the summer tropic.

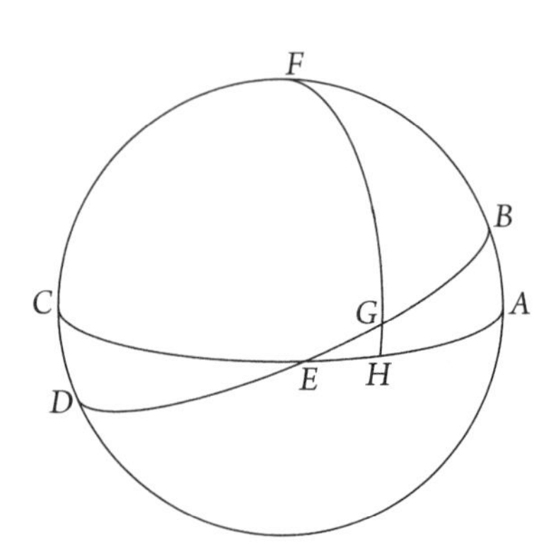

H77

And let there be taken on arc *ABC* the pole of the equatorial circle *AEC* and let it be point *F*, and let arc *EG* of the mid-zodiac circle, assumed as 30 parts of which a great circle is 360, be intercepted.

And through *F* and *G* let there be described the arc of a great circle, *FGH*, and let it be proposed to find the arc, namely *GH*.

Now, let it have been previously assumed here and in general for all the like proofs, in order that we not say the same things of each, that, when we speak of the magnitudes of arcs or chords, of so many degrees or parts, in the case of the arcs, we say they are such as the circumference of a great circle is 360 parts, while in the case of chords such as the diameter of a circle is 120 parts.

Therefore, since in a diagram of great circles, two arcs *FH* and *EB* have been drawn to two arcs *AF* and *AE*, cutting each other at *G*,
the ratio of the chord subtending the arc double the arc *FA* to the chord subtending the arc double the arc *AB* is compounded from the ratio of the chord subtending the arc double the arc *HF* to the chord subtending the arc

[1] *the plane diagram of straight lines*: that is, the diagram for Lemma 1.3.

double the arc *HG* and the ratio of the chord subtending the arc double the arc *GE* to the chord subtending the arc double the arc *EB* {H76.3}.

But the arc double the arc *FA* is 180 degrees and the chord subtending it is 120 parts,
while the arc double the arc *AB* is 47;42,40 degrees, according to the ratio that we agreed to, of 83 to 11 {H68.4},
while the chord subtending it is 48;31,55 parts.

H78 And, again, the arc double the arc *GE* is 60 degrees and the chord subtending it 60 parts,
while the arc double the arc *EB* is 180 degrees and the chord subtending it is 120 parts.

Therefore, if we divide out the ratio of 60 to 120 from the ratio of 120 to 48;31,55,
there remains the ratio of the chord that subtends the arc double the arc *FH* to the chord subtending the arc double the arc *HG*:
that of 120 to 24;15,57.

And the arc double the arc *FH* is 180 degrees, while the chord subtending it is 120 parts.

And therefore the chord subtending the arc double the arc *HG* is 24;15,57 of the same.

So that also, the arc double the arc *HG* is 23;19,59 degrees, and the arc *HG* itself is 11;40, most nearly, of the same.

Calculation 1.31

Again, let arc *EG* be assumed as 60 degrees, so that, the rest remaining the same, the arc double the arc *EG* is 120 degrees, while the chord subtending it is 103;55,23 parts.

Therefore, if, again, we divide out the ratio of 103;55,23 to 120 from the ratio of 120 to 48;31,55,
there will be left the ratio of the chord subtending the arc double the arc *FH* to the chord subtending the arc double the arc *HG*:
that of 120 to 42;1,48.

And the chord subtending the arc double the arc *FH* is 120 parts;
so that the chord subtending the arc double the arc *HG* also will be 42;1,48 of the same.

And therefore the arc double the arc *HG* is 41;0,18 degrees,
while arc *HG* is 20;30,9 of the same:
which it was necessary to prove.

H79 Now, by calculating in the same way the magnitudes for particular arcs as well, we will set out a table of the 90 degrees of the quadrant, which contains side by side magnitudes of the arcs similar to those that have been proven. And the table is as follows.

15. Table of Obliquity.

H80

arcs of the mid-zodiac circle	arcs of the meridian circle		
1	0	24	16
2	0	48	31
3	1	12	46
4	1	37	0
5	2	1	12
6	2	25	22
7	2	49	30
8	3	13	35
9	3	37	37
10	4	1	38
11	4	25	32
12	4	49	24
13	5	13	11
14	5	36	53
15	6	0	31
16	6	24	1
17	6	47	26
18	7	10	45
19	7	33	57
20	7	57	3
21	8	20	0
22	8	42	50
23	9	5	32
24	9	28	5
25	9	50	29
26	10	12	46
27	10	34	57
28	10	56	44
29	11	18	25
30	11	39	59
31	12	1	20
32	12	22	30
33	12	43	28
34	13	4	14
35	13	24	47
36	13	45	6
37	14	5	11

arcs of the mid-zodiac circle	arcs of the meridian circle		
38	14	25	2
39	14	44	39
40	15	4	4
41	15	23	10
42	15	42	2
43	16	0	38
44	16	18	58
45	16	37	20
46	16	54	47
47	17	12	16
48	17	29	27
49	17	46	20
50	18	2	53
51	18	19	15
52	18	35	5
53	18	50	41
54	19	5	57
55	19	20	56
56	19	35	28
57	19	49	42
58	20	3	31
59	20	17	4
60	20	30	9
61	20	42	58
62	20	55	24
63	21	7	21
64	21	18	58
65	21	30	11
66	21	41	0
67	21	51	25
68	22	1	25
69	22	11	11
70	22	20	11
71	22	28	57
72	22	37	17
73	22	45	11
74	22	52	39
75	22	59	41

arcs of the mid-zodiac circle	arcs of the meridian circle		
76	23	6	17
77	23	12	27
78	23	18	11
79	23	23	28
80	23	28	16
81	23	32	30
82	23	36	35
83	23	40	2
84	23	43	2
85	23	45	34
86	23	47	39
87	23	49	16
88	23	50	25
89	23	51	6
90	23	51	20

H81

16. On Ascensions in the Right Sphere.

H82.2

The next task would be to demonstrate as well the magnitudes of the arcs of the equatorial circle produced by circles described through its poles and the given segments of the oblique circle. For in this way, we will have in how many equatorial time-degrees[1] the segments of the mid-zodiac circle will invariably traverse both the meridian circle and the horizon in upright sphere; because at that time alone it is described through the poles of the equatorial circle.

Proposition 1.9
Calculation 1.32

Therefore, let there be set out the previously shown diagram. And, when arc *EG* of the oblique circle is again first given as 30 parts, let it be required to find arc *EH* of the equatorial circle.

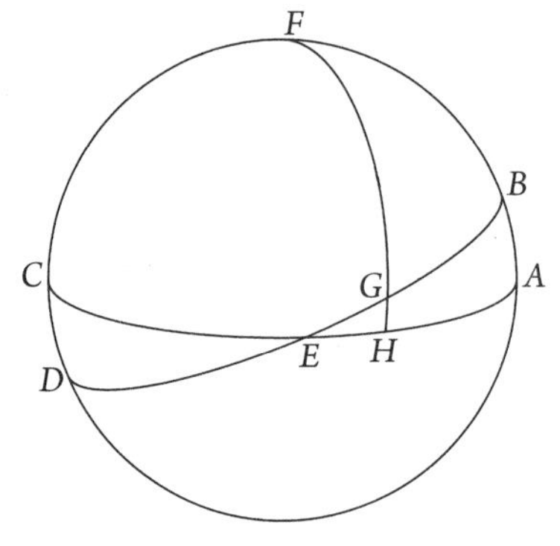

Now, by the same considerations as in the preceding, the ratio of the chord subtending the arc double the arc *FB* to the chord subtending the arc double the arc *BA* is compounded from the ratio of the chord subtending the arc double the arc *FG* to the chord subtending the

1 *equatorial time-degrees*: that is, degrees on the equator. Each such degree is traversed in the same interval of time by the Motion of the Same (see Preliminaries, page 16 above). Ptolemy will use the term "time-degrees" repeatedly in subsequent discussions.

arc double the arc *GH* and the ratio of the chord subtending the arc double the arc *HE* to the chord subtending the arc double the arc *EA* {H74.15}.

But the arc double the arc *FB* is 132;17,20 degrees and the chord subtending it is 109;44,53 parts,

H83 while the arc double the arc *BA* is 47;42,40 degrees and the chord subtending it is 48;31,55 parts.

And, again, the arc double the arc *FG* is 156;40,1 degrees and the chord subtending it is 117;31,15 parts,

while the arc double the arc *GH* is 23;19,59 degrees and the chord subtending it is 24;15,57 parts.

Therefore, if we divide out the ratio of 117;31,15 to 24;15,57 from the ratio of 109;44,53 to 48;31,55,

there will be left for us the ratio of the chord subtending the arc double the arc *HE* to the chord subtending the arc double the arc *EA*:

that of 54;52,26 to 117;31,15;

and 56;1,25 to 120 also is the same ratio.

And the arc double the arc *EA* is 180 degrees, while the chord subtending it is 120 parts.

And therefore the chord subtending the arc double the arc *HE* is 56;1,25 of the same parts.

So that also, the arc double the arc *HE* will be 55;40 degrees, most nearly, while arc *HE* will be 27;50 of the same.

Again let arc *EG* be assumed as 60 degrees, Calculation 1.33

so that, the rest remaining the same, the arc double the arc *FG* is 138;59,42 degrees and the chord subtending it 112;23,56 parts,

and the arc double the arc *GH* 41;0,18 degrees and the chord subtending it 42;1,48 parts.

Therefore, if we divide out the ratio of 112;23,56 to 42;1,48 from the ratio of 109;44,53 to 48;31,55,

there will be left the ratio of the chord subtending the arc double the arc *HE*

H84 to the chord subtending the arc double the arc *EA*:

that of 95;2,40 to 112;23,56;

the ratio of 101;28,20 to 120 also is the same as this.

And the chord subtending the arc double the arc *EA* is 120 parts;

so that also, the chord subtending the arc double the arc *HE* is 101;28,20 of the same.

And therefore the arc double the arc *HE* will be 115;28 degrees, most nearly,

while arc *HE* itself will be 57;44 of the same.

And it has been shown that the first sign of the mid-zodiac circle, from the equinoctial point, rises synchronously with 27;50 parts of the equatorial circle in the manner being set out, Calculation 1.34

while the second rises synchronously with 29;54 parts,
since both together were proved as 57;44 degrees.

And the third sign will clearly rise synchronously with the 32;16 degrees supplementary to the quadrant;
because the whole quadrant of the oblique circle also rises synchronously with the whole quadrant of the equatorial circle, in relation to circles described through the poles of the equatorial circle.

Now, pursuing the present demonstration in the same way, we also calculated the arcs of the equatorial circle that rise synchronously with each ten-degree span of the oblique circle, because arcs of still smaller fractions than these differ in nothing significant from the differences due to uniform augmentation. Accordingly, we will set these out also,[1] in order that we have ready at hand in how many time-degrees each of them will invariably traverse the meridian circle, as we said {H82.6–9}, and the horizon in right sphere. We will make our starting point the ten-degree span at the equinoctial point.

H85

Calculations 1.34 – 1.44

The first, then, contains 9;10 time-degrees,
the second 9;15 time-degrees,
and the third 9;25 time-degrees;
so that the time-degrees of the first sign of the zodiac together sum up to 27;50 time-degrees.

The fourth contains 9;40 time-degrees,
the fifth 9;58 time-degrees,
and the sixth 10;16 time-degrees;
so that also the time-degrees of the second sign of the zodiac sum up to 29;54 time-degrees.

The seventh contains 10;34 time-degrees,
the eighth 10;47 time-degrees,
and the ninth 10;55 time-degrees;
so that, again, the time-degrees of the third sign of the zodiac, also at the tropic points, sum up to 32;16 time-degrees.

And the time-degrees of the whole quadrant, in good agreement, sum up to 90 time-degrees.

And it is immediately clear that the ordering in the remaining quadrants too is actually the same. For everything results the same in each quadrant, because right sphere is assumed; that is to say, the equatorial circle at right angles to the horizon.

---

1 *we will set these out also*: Ptolemy's table in Book II, Chapter 8 is omitted in this edition.

Book II

H87

1. On the General Position of the Zone Inhabited by Us.

In the first book of our work, we explained what ought to be assumed first concerning the disposition of all the heavenly bodies[1] in summary fashion and everything that one might deem useful in the right sphere for the investigation of our subject. We will attempt next to treat methodically as well of the salient occurrences in the inclined sphere; again, to the greatest extent possible, in the most manageable way.

Now, what ought in general to be assumed here first is the following: when H88 the earth is divided into four quadrants, which are produced by the plane of the equatorial circle[2] and by one of the planes described through its poles, the magnitude of the zone inhabited by us is contained, most nearly, by one of the northern quadrants. And this would prove most clear because in the case of latitude, that is to say, the passage from south to north, the meridian shadows of gnomons on the equinoxes everywhere make their inclinations towards the north and never towards the south. And, in the case of longitude, that is to say, the passage from east to west, the same eclipses, especially the lunar ones, are observed by those living at the easternmost extremities of the zone inhabited by us and by those at the westernmost extremities about the same time, no earlier or later than twelve equinoctial hours in longitude of the quadrant that contains a twelve hour interval,[3] since it is bounded by one of the semicircles of the equatorial circle.

As for what ought in particular to be investigated, one might suppose what is most suitable for the present treatment is the particulars of the circles more northerly than the equatorial circle, parallel to it and to the assumed zones of habitation, in terms of their more salient, distinctive features. These H89 are: ⟨1⟩ by how much the poles of the primary movement[4] are removed from the horizon, or by how much the vertex-point of the equatorial circle[5]

---

1 *all the heavenly bodies*: περὶ τῆς τῶν ὅλων σχέσεως.

2 *the plane of the equatorial circle*: τοῦ κατὰ τὸν ἰσημερινὸν κύκλον. Theon (II. 604.20ff) supplies τὸ ἐπίπεδον here.

3 *the quadrant that contains a twelve hour interval*: Note that 12 equinoctial hours represents *half* the circumference of the equator, and therefore bounds a *quadrant* of the earth's surface.

4 *poles of the primary movement*: that is, the poles of the equator.

5 *the vertex-point of the equatorial circle*: the highest point on the equator, that is, where it crosses the meridian. The distance (on the meridian) between a pole of the equator and the horizon equals that between the equator and the zenith..

is removed on the meridian circle; and ⟨2⟩ for which observers[1] the sun is at its vertex,[2] when and how often this sort of thing results; and ⟨3⟩ what ratios the equinoctial and tropic shadows at mid-days have to ⟨their⟩ gnomons; and ⟨4⟩ how great are the differences of the longest or shortest days compared with equinoctial days; and ⟨5⟩ all the other things that are observed concerning the particular augmentations of days and nights; and, furthermore, ⟨6⟩ concerning the co-risings and co-settings of the equatorial and oblique circles; and ⟨7⟩ concerning the additional distinctive properties and magnitudes of the angles produced by the principal great circles.

2. How, When the Magnitude of the Longest Day is Given, the Arcs of the Horizon Intercepted by the Equatorial and Oblique Circles are Given.

Proposition 2.1
Calculation 2.1

Now, in general, by way of examples, let there be proposed the circle described through Rhodes, parallel to the equatorial circle, where the elevation of the pole is 36 degrees and the longest day is 14½ equinoctial hours.[3]

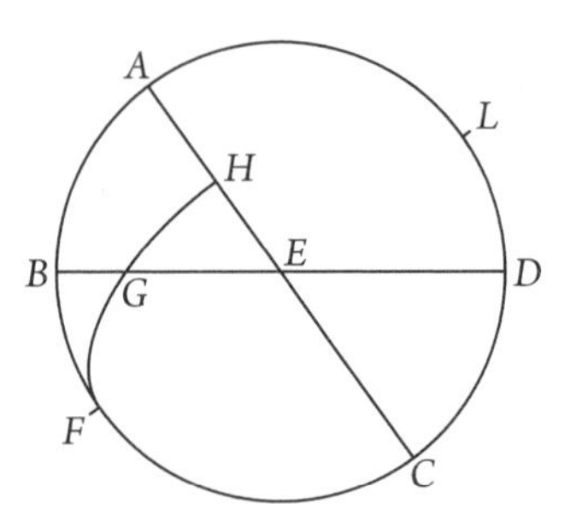

H90

And let *ABCD* be a meridian circle, *BED* the eastern semicircle of the horizon, and, similarly, *AEC* the semicircle of the equatorial circle, and *F* its southern pole. Let the winter tropic point of the mid-zodiac circle be assumed to rise through *G*, and let quadrant *FGH* of a great circle be described through *F* and *G*.

First, let the magnitude of the longest day be given, and let it be proposed to find arc *EG* of the horizon.

Therefore, since the rotation of the sphere is produced around the poles of the equatorial circle, it is clear that point *G* and point *H* will be on the meridian circle *ABCD* at the same time, and that the time-span from rising up to the culmination of *G* above the earth is contained by arc *HA* of the equatorial circle, while the time-span from the culmination under the earth up to rising is contained by arc *CH*. It follows that the time-span of the day also is twice the time contained by *HA*, while that of the night is twice the time-span contained by *CH*, since the segments above the earth and below

H91

1 *for which observers*: οἷς

2 *the sun is at its vertex*: Here, "the sun reaches the zenith."

3 *equinoctial hours*: One equinoctial hour corresponds to 15 degrees along the equator, for the sun lies on the equator at equinox..

the earth of all the circles parallel to the equatorial circle are bisected by the meridian circle quite independently.

Proposition 2.2
Calculation 2.2

For this reason, arc *EH* also, being half of the difference of the shortest or longest day compared with the equinoctial day, is 1¼ hour on the assumed parallel, and, evidently, 18;45 time-degrees,[1]
while the supplement to the quadrant, *HA*, is 71;15 of the same.

Accordingly, since in the same way as what was previously demonstrated,[2] two arcs *EB* and *FH* have been drawn to two arcs *AE* and *AF* of great circles and cut each other at *G*,
the ratio of the chord subtending double the arc *HA* to the chord subtending the arc double the arc *AE* is compounded from the ratio of the chord subtending the arc double the arc *HF* to the chord subtending the arc double the arc *FG* and the ratio of the chord subtending the arc double the arc *GB* to the chord subtending double arc *BE* {H76.3}.

But the arc double the arc *HA* is 142;30 degrees and the chord subtending it is 113;37,54 parts,

H92

while the arc double the arc *AE* is 180 degrees and the chord subtending it is 120 parts.

And, again, the arc double the arc *HF* is 180 degrees and the chord subtending it is 120 parts,
while the arc double the arc *FG* is 132;17,20 degrees and the chord subtending it is 109;44,53 parts.

Therefore, if we divide out[3] the ratio of 120 to 109;44,53 from the ratio of 113;37,54 to 120,
there will remain for us the ratio of the chord subtending the arc double the arc *GB* to the chord subtending the arc double the arc *BE*,
that of 103;55,23 to 120.

And the chord subtending the arc double the arc *BE*, since it occupies a quadrant, is 120 parts.

And therefore the chord subtending the arc double the arc *GB* also is 103;55,23 of the same.

So that also, the arc double the arc *BG* will be 120 degrees, most nearly, while arc *BG* itself will be 60 of the same.

And therefore the remainder *GE* is left as 30 degrees of which the horizon is 360: which it was necessary to prove.

---

1 1¼ *hour*: Half the difference between 14½ hours, the longest day at Rhodes {H90} and the equinoctial day to 12 hours.

2 *what was previously demonstrated*: That is, the (spherical) Menelaus Theorems, Props. 1.6 and 1.7. See also Appendix III.

3 *divide out*: Here and subsequently, *αφέλωμεν*. To "divide out the ratio" means to remove it from a compound of ratios.

3. How, When the Same Things are Assumed, the Elevation of the Pole is Given, and Conversely.

Proposition 2.3
Calculation 2.3

Now, let it be proposed again, when this is given to find the elevation of the pole also, that is to say arc *BF* of the meridian circle.

Therefore, in the same diagram, the ratio of the chord subtending the arc double the arc *EH* to the chord subtending the arc double the arc *HA* is compounded from the ratio of the chord subtending the arc double the arc *EG* to the chord subtending the arc double the arc *GB* and the ratio of the chord subtending the arc double the arc *BF* to the chord subtending the arc double the arc *FA* {H74.15}.

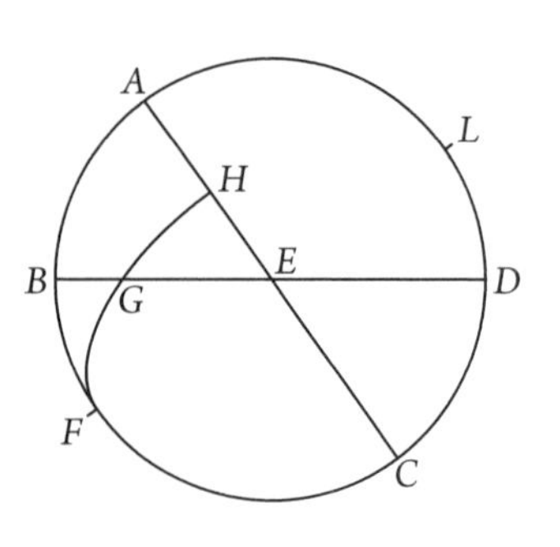

H93

But the arc double the arc *EH* is 37;30 degrees and the chord subtending it is 38;34,22 parts,
while the arc double the arc *HA* is 142;30 degrees and the chord subtending it is 113;37,54 parts.

And, again, the arc double the arc *EG* is 60 degrees and the chord subtending it is 60 parts,
while the arc double the arc *GB* is 120 degrees and the chord subtending it is 103;55,23 parts.

Therefore, if we divide out the ratio of 60 to 103;55,23 from the ratio of 38;34,22 to 113;37,54
there will be left the ratio of the chord subtending the arc double the arc *BF* to the chord subtending the arc double the arc *FA*,
that of 70;33, most nearly, to 120.

And again the chord subtending the arc double the arc *FA* is 120 parts; therefore the chord subtending the arc double the arc *BF* is 70;33 of the same parts;
so that the arc double the arc *BF* will be 72;1 degrees, and the arc *BF* will be 36 of the same degrees, most nearly.

Proposition 2.4
Calculation 2.4

Again, in the same diagram, conversely, let arc *BF* of the elevation of the pole be given, observed as 36 degrees,
and let it be proposed to find the difference of the shortest or longest day compared with the equinoctial day,
that is to say the arc double the arc *EH*.

H94

Therefore, through the same considerations {H74.15}, the ratio of the chord subtending the arc double the arc *FB* to the chord subtending the arc double the arc *BA* is compounded from the ratio of the chord subtending the arc double the arc *FG* to the chord subtending the arc double the arc *GH* and the ratio of the chord subtending the arc double the arc *HE* to the chord subtending the arc double arc *EA*.

But the arc double the arc *FB* is 72 degrees and the chord subtending it is 70;32,3 parts,
while the arc double the arc *BA* is 108 degrees and the chord subtending it is 97;4,56 parts.

And, again, the arc double the arc *FG* is 132;17,20 degrees and the chord subtending it is 109;44,53 parts,
while the arc double the arc *GH* is 47;42,40 degrees and the chord subtending it is 48;31,55 parts.

Therefore, if we divide out the ratio of 109;44,53 to 48;31,55 from the ratio of 70;32,3 to 97;4,56,
there will remain for us the ratio of the chord subtending the arc double the arc *HE* to the chord subtending the arc double the arc *EA*,
that of 31;11,23 to 97;4,56.

And since 38;34 to 120 has the same ratio, most nearly,
H95 while the chord subtending the arc double the arc *EA* is 120 parts,
the chord subtending the arc double the arc *EH* also is concluded as 38;34 of the same.

So that also, the arc double the arc *EH* will be 37;30 degrees, most nearly, and 2½ equinoctial hours; which it was necessary to prove.

Proposition 2.5
Calculation 2.5

In the same way, arc *EG* of the horizon will also be given, because the ratio of the chord subtending the arc double the arc *FA* to the chord subtending the arc double the arc *AB*, being given {H76.3},
also is compounded from the ratio of the chord subtending the arc double the arc *FH* to the chord subtending the arc double the arc *HG*, itself given also,
and from the ratio of the chord subtending the arc double the arc *GE* to the chord subtending the arc double the arc *EB*.

So that also, when *EB* is given, the magnitude of EG also remains.

Proposition 2.6
Calculation 2.6

It is clear that, even if we do not assume the winter tropic point as G, but any of the remaining segments of the mid-zodiac circle,[1] in the same way, again, each of the arcs *EH* and *EG* will be given. We set out, through the table of obliquity, arcs of the meridian circle that are intercepted by each segment of the mid-zodiac circle and the equatorial circle; that is to say, arcs similar
H96 to arc *GH*. And it immediately follows that the segments of the mid-zodiac circle that are produced by the same parallels, that is to say, those equally removed from the same tropic point, make the sections of the horizon the same and always in the same direction of the equatorial circle, and make the

[1] *even if we do not assume the winter tropic point as G*: That is, even if we take as given *any* segment of the ecliptic, not necessarily one bordering the tropic.

magnitudes of days and nights equal, each like to like. And it is demonstrated as well that the segments that are produced by equal parallels, that is to say, those equally removed from the same equinoctial point, make the arcs of the horizon equal, from each side of the equatorial circle, and make the magnitudes of days and nights alternately equal, each unlike to unlike.[1]

Proposition 2.7
Calculation 2.7

For if, in the diagram that is set out, we assume also point *I*, at which the circle equal and parallel to that described through *G* cuts the semicircle of the horizon, *BED*, and we complete segments *GJ* and *IK* of parallel circles that are evidently alternate and equal and we describe quadrant *LIM* through *I* and the northern pole,

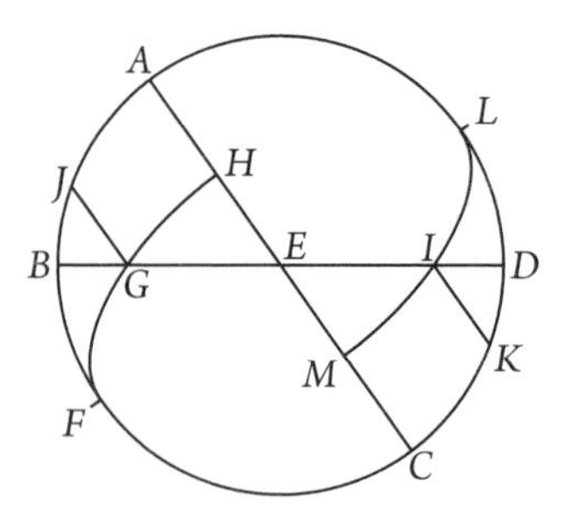

arc *HA* will be equal to *MC*,
because *JG* and *IK* are each similar to the other,
and the remaining arc *EH* also will be left equal to the remaining arc *EM*.

And in two similar triangles, *EGH* and *EIM*, two sides will equal two sides, *EH* to *EM*, and *GH* to *IM*; H97
and each of the angles at *H* and *M* is right.

So that base *EG* also is equal to base *IE*.

[The rest of Book II is omitted]

1 τῶν ὁμοίων (like to like); τῶν ἀνομοίων (unlike to unlike). Theon II. 629.2–3 has for the latter, "That is, a night to a day, and a day to a night"; so the former must mean "a night to a night, and a day."

PRELIMINARIES TO BOOK III

Precession of the Equinoxes

In the first chapter of Book III, Ptolemy will argue that the sun's annual period (that is, its completion of a full eastward longitudinal cycle) should be determined by its return to the same solstitial or equinoctial point rather than by its return to the same position among the zodiac constellations. The reason for this, he says, is that the sphere of the fixed stars makes, in addition to its westward diurnal motion, a small eastward movement that only becomes apparent over a long period of time. Ptolemy will calculate this motion in Book VII and will show there that it takes place about the axis of the zodiac.

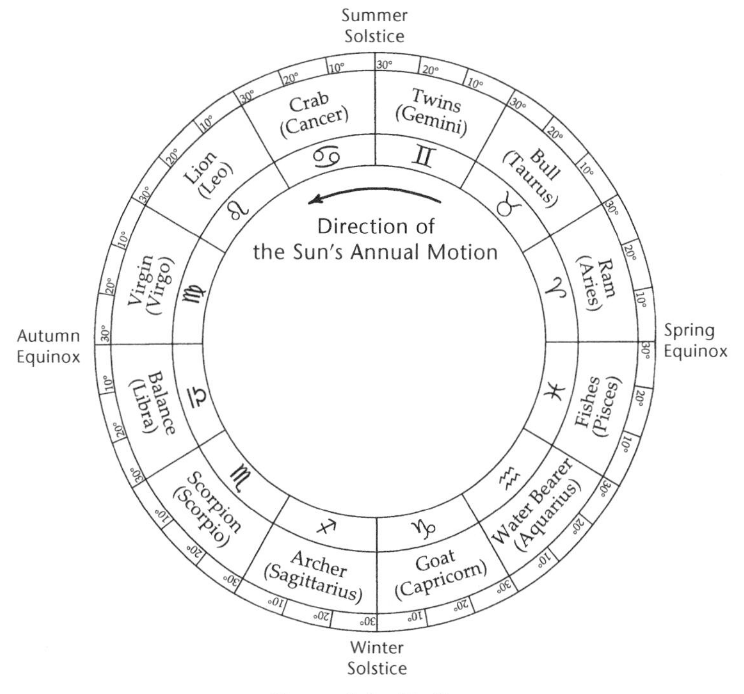

Signs of the Zodiac

By definition, the sequence of zodiac signs begins with the spring equinox; and since in Ptolemy's time the first thirty-degree sector of the zodiac contained the constellation Ram (Aries), the first of the zodiac signs was itself named Ram. We continue to call it Ram today; but because the zodiac constellations have been slipping, as it were, in the eastward direction, the constellation that presently occupies the sign Ram is Fishes (Pisces). If we expected the spring equinox to coincide with the constellation Ram, but observed instead that it actually coincides with the constellation Fishes, we might say that equinox occurred *too soon*—that it *preceded* its expected time. Hence the name "precession of the equinoxes."

The following table shows the relationship between the zodiac constellations and signs in Ptolemy's time and in ours.[1]

Zodiac Sign	Constellation in Antiquity	Current Constellation	Dates Occupied by Sun
Ram	Aries	Pisces	Mar 21–Apr 20
Bull	Taurus	Aries	Apr 20–May 21
Twins	Gemini	Taurus	May 21–Jun 22
Crab	Cancer	Gemini	Jun 22–Jul 23
Lion	Leo	Cancer	Jul 23–Aug 24
Virgin	Virgo	Leo	Aug 24–Sept 23
Balance	Libra	Virgo	Sept 23–Oct 24
Scorpion	Scorpio	Libra	Oct 24–Nov 23
Archer	Sagittarius	Scorpio	Nov 23–Dec 22
Goat	Capricorn	Sagittarius	Dec 22–Jan 20
Water-bearer	Aquarius	Capricorn	Jan 20–Feb 19
Fishes	Pisces	Aquarius	Feb 19–Mar 21

Mean Sun and Apparent Sun

For many purposes it is sufficient to treat the sun as though it moved uniformly about the ecliptic. But in Book III, Chapter 4, Ptolemy will note that the number of days observed between successive equinoxes and solstices is not exactly one-quarter of a year, as we should then expect. Instead, it varies between about 88 and about 94 days. The visible, or *apparent*, sun does not, therefore, move uniformly about the ecliptic. But it will often prove very useful to compare a given position of the apparent sun to where it would be observed if it *did* exhibit regular, uniform motion. A hypothetical "sun"—a mathematical construct that completes one revolution about the ecliptic in the same time as does the apparent sun, but does so with uniform motion—has come to be called the *mean sun*. Ptolemy employs the concept of a mean sun but does not use that name.

In the table that follows, the positions of the apparent sun in the zodiac, together with the incremental amounts of the sun's motion eastward, are recorded at 5-day intervals for the year 2010. The inequalities in its motion are small but evident; they are comparable to what Ptolemy might have observed in his day. Although we do not possess any records of Ptolemy's solar observations, his argument at the beginning of Chapter 4 implies that he had access to such data.

[1] Adapted from Evans, James, *The History and Practice of Ancient Astronomy*, Oxford University Press (1988).

Eastward Motion of the Apparent Sun through the Zodiac, 2010

Date	Position	Difference	Date	Position	Difference
Jan 1	Goat 10°27′		SUMMER SOLSTICE Crab 0		
Jan 6	Goat 15 32	5°05′	Jun 25	Crab 3°21′	4°46′
Jan 11	Goat 20 38	5 06	Jun 30	Crab 8 07	4 46
Jan 16	Goat 25 44	5 06	Jul 5	Crab 12 53	4 46
Jan 21	Water Bearer 0 49	5 05	Jul 10	Crab 17 39	4 46
Jan 26	Water Bearer 5 55	5 06	Jul 15	Crab 22 26	4 47
Jan 31	Water Bearer 10 59	5 04	Jul 20	Crab 27 12	4 46
Feb 5	Water Bearer 16 04	5 05	Jul 25	Lion 1 58	4 47
Feb 10	Water Bearer 21 08	5 04	Jul 30	Lion 6 45	4 47
Feb 15	Water Bearer 26 11	5 03	Aug 4	Lion 11 32	4 47
Feb 20	Fishes 1 14	5 03	Aug 9	Lion 16 19	4 47
Feb 25	Fishes 6 16	5 02	Aug 14	Lion 21 07	4 48
Mar 2	Fishes 11 17	5 01	Aug 19	Lion 25 56	4 49
Mar 7	Fishes 16 18	5 01	Aug 24	Virgin 0 44	4 48
Mar 12	Fishes 21 18	5 00	Aug 29	Virgin 5 34	4 50
Mar 17	Fishes 26 17	4 59	Sep 3	Virgin 10 24	4 50
SPRING EQUINOX Ram 0			Sep 8	Virgin 15 15	4 51
Mar 22	Ram 1 15	4 58	Sep 13	Virgin 20 07	4 52
Mar 27	Ram 6 13	4 58	Sep 18	Virgin 24 59	4 52
Apr 1	Ram 11 09	4 56	Sep 23	Virgin 29 52	4 53
Apr 6	Ram 16 05	4 56	AUTUMN EQUINOX Balance 0		
Apr 11	Ram 21 00	4 55	Sep 28	Balance 4 46	4 54
Apr 16	Ram 25 54	4 54	Oct 3	Balance 9 41	4 55
Apr 21	Bull 0 47	4 53	Oct 8	Balance 14 36	4 55
Apr 26	Bull 5 40	4 53	Oct 13	Balance 19 33	4 57
May 1	Bull 10 31	4 51	Oct 18	Balance 24 30	4 57
May 6	Bull 15 22	4 51	Oct 23	Balance 29 28	4 58
May 11	Bull 20 12	4 50	Oct 28	Scorpion 4 27	4 59
May 16	Bull 25 02	4 50	Nov 2	Scorpion 9 27	5 00
May 21	Bull 29 51	4 49	Nov 7	Scorpion 14 28	5 01
May 26	Twins 4 39	4 48	Nov 12	Scorpion 19 29	5 01
May 31	Twins 9 27	4 48	Nov 17	Scorpion 24 31	5 02
Jun 5	Twins 14 15	4 48	Nov 22	Scorpion 29 34	5 03
Jun 10	Twins 19 02	4 47	Nov 27	Archer 4 37	5 03
Jun 15	Twins 23 49	4 47	Dec 2	Archer 9 41	5 04
Jun 20	Twins 28 35	4 46	Dec 7	Archer 14 45	5 04

Date	Position	Difference
Dec 12	Archer 19°50′	5°05′
Dec 17	Archer 24 55	5 05
WINTER SOLSTICE Goat 0		
Dec 22	Goat 0 01	5 06
Dec 27	Goat 5 06	5 05

Date	Position	Difference
2011 Jan 1	Goat 10°12′	5°06′
Jan 6	Goat 15 18	5 06
Jan 11	Goat 20 24	5 06
Jan 16	Goat 25 29	5 05

Dates of Equinoxes and Solstices

In the preceding table, Spring Equinox falls between Mar 17 and Mar 22, 2010, during which time the apparent sun moves 4°58′. But the apparent sun needs to move only 3°43′ to carry it from Fishes 26°17′ (its position on Mar 17) to Spring Equinox at Ram 0°. At the rate of 4°58′ per five days, therefore, the apparent sun will reach equinox about $3\frac{3}{4}$ days after Mar 17, that is, on Mar 20, 2010. By similar interpolations we may find Summer Solstice on Jun 21, Autumn Equinox on Sep 23, and Winter Solstice on Dec 21, 2010.

The interval from Spring Equinox to Summer Solstice is thus about 93 days,[1] that from Summer Solstice to Autumn Equinox about 94 days, that from Autumn equinox to Winter Solstice about 89 days, leaving the time from Winter Solstice to the next Spring Equinox as about 89 days. Ptolemy will make use of similar day counts in Chapter 4.

Greatest, Mean, and Least Motion

In the preceding table, notice that from about Dec 27 to about Jan 16 the apparent sun moves at the rate of 5°6′ per five-day interval—its greatest motion over the course of the year. Since the values have been estimated to the nearest minute, it is reasonable to suppose that the apparent sun actually achieves its greatest motion midway through this period, or about Jan 6.

Similarly, the apparent sun moves at its slowest rate, 4°46′ per five-day interval, from about Jun 17 to about Jul 23; we may surmise that the apparent sun actually achieves its least motion midway through this period, or about Jul 5.

In Chapter 1 below, Ptolemy will determine the length of the year as about 365; 14, 48 days, that is, 365.2466 days; thus the sun's mean motion is 360°÷365.2466 or about 0°59′ per day, or 4°56′ per five-day interval. In the preceding table it appears that the apparent sun moves at this rate on about Apr 3 and again about Oct 10.

Thus the apparent sun's time from greatest to mean motion is from Jan 6 to Apr 3, or 87 days. This is less than its time from mean to least motion, which is from Apr 3 to Jul 5, or 93 days. In Chapter 3, Ptolemy will shape his hypotheses in accordance with this inequality.

[1] For this and similar day counts Appendix 2, *Days of the Year*, will be useful.

Book III

⟨Introduction⟩

In the preceding books {I–II} we have explained mathematically what H191 ought generally to be premised concerning the heavens and the earth, and, furthermore, the inclination of the sun's circle through the middle of the zodiac signs, and what in particular results for this circle, both in right sphere and oblique sphere at each zone of habitation.[1] We think it follows directly after these matters to give an account of the sun and the moon and to explain in detail what additionally results concerning their motions, since none of the appearances relating to the stars[2] can be at all discovered without preliminary explanation of these matters. And we find as foremost among these matters the treatment of the sun's motion, without which, again, it would not be possible to apprehend in detail what relates to the moon.

1. On the Magnitude of the Annual Period.

Now, to find the annual period is the first task among all the things demonstrated about the sun. We may learn the disagreements and perplexities of the ancients in articulating this period from what has been written by them and especially by Hipparchus, a man who loved labor and, at the same time, truth. For the fact that the annual period, when determined through apparent periodic returns at the tropics and equinoxes, is found shorter than H192 $365\frac{1}{4}$ days, but, when determined through observations of the fixed stars, is found longer most of all leads him into such perplexity. Hence he focuses on the fact that the sphere of the fixed stars itself also makes some movement[3] over a long time, just as do the spheres of the planets, eastwards of the movement that produces the ⟨diurnal westward⟩ primary revolution, on the circle that is described through the poles of both the equatorial circle and the oblique circle. We will demonstrate that this is the case and how it happens in our books on the fixed stars {VII–VIII}; for what relates to those stars could not be investigated at all without the preliminary explanation of the sun and

---

[1] *each zone of habitation*: that is, each latitude.

[2] *the stars*: Ptolemy uses the word "star" not only for the fixed stars but for the "wandering stars" (the sun, moon, and five planets).

[3] *the sphere of the fixed stars itself also makes some movement*: A reference to the precession of the equinoxes, mentioned in the Preliminaries to Book III; Ptolemy will discuss it in Book VII.

moon. But for the present enquiry, we think that we ought to inquire into the annual period of the sun with nothing else in view than the periodic return of the sun itself upon itself; that is to say, upon the oblique circle that the sun produces.[1] And we think that we ought to define the annual period as that during which, from a certain unmoved point on this circle, it next arrives at the same point. We deem as proper starting points of this periodic return only the points of the previously mentioned circle that are determined by the tropic and equinoctial points. For if we focus mathematically on the account, neither will we find a more proper periodic return than that which carries the sun to the same configuration both in terms of place and time (this return is observed either relative to the horizons, or to the meridian circle, or to the magnitudes of days and nights), nor will we find other starting points in the mid-zodiac circle, but only those that are incidentally determined by the tropic and the equinoctial points. And if someone inquires in a more physical way into what is appropriate, neither will he find a periodic return that is more reasonable than that which carries the sun from the same weather to the same, and from the same season to the same. Nor will he find other starting points than those alone, by which above all the seasons are distinguished. Furthermore it appears absurd to use the periodic return that is observed relative to the fixed stars both for other reasons and especially because their sphere is observed performing a certain regular movement eastwards in the heavens. For, were this the case, nothing will prevent our saying that the annual period of the sun is the amount of time in which, for example, the sun overtakes the star Saturn or one of the other planets. And in this way there would be many different kinds of annual periods. For these reasons, we think it appropriate that this be considered the annual period of the sun: the period found through observations taken, to the greatest extent possible, over a rather long interval, from some tropic or equinox to the next identical one. **H193** **H194**

Since the inequality of even this sort of periodic return, which Hipparchus came to suspect through continuous successive observations, alarmed him to some extent, we will attempt briefly to show that this is by no means cause for alarm. We became confident that these periods are not unequal on the basis of the tropics and equinoxes that we ourselves have actually successively observed using instruments. For we find that the periods differ from the $\frac{1}{4}$ day surplus to no significant extent, except sometimes roughly with as great an error as admits of being due to the construction and position of the instruments. And we also infer from those values, on the basis of which Hipparchus makes his calculations, that the error regarding the inequalities lies rather with the observations.

[1] *the oblique circle that the sun produces*: the mid-zodiac circle or ecliptic.

For after he first sets out in his *On the Change of the Tropic and Equinoctial Points* the summer and winter tropics that seem to him to have been observed accurately and successively, he himself also agrees that the inconsistency in them is not so great that one can discern the inequality of the annual period through them. For he adds to them the following:

> From these observations, then, it is clear that the differences among the annual periods are very small indeed. But in the case of the tropics I am confident that we and Archimedes, both in observing and calculating, err up to a quarter of a day. The anomaly of annual periods can be accurately apprehended from what has been observed on the bronze armillary sphere, situated in Alexandria in the so-called Square Porch, which seems clearly to indicate the equinoctial day as that on which it begins to become illuminated on its concave surface from the other direction.

H195

Observations 3.1 – 3.6, Calculations 3.1 – 3.5

Then he lays out, first, dates of autumnal equinoxes, observed with the highest possible accuracy. One occurred in the 17th year of the third Calippic cycle, Mesore 30, about the setting of the sun, and one 3 years later in the 20th year on the first intercalated day, early morning; it should have occurred at noon, so that there is a discrepancy of $\frac{1}{4}$ of one day. A year later, one occurred in the 21st year at the 6th hour, which in fact was also in agreement with the observation before it. Eleven years later, one occurred in the 32nd year on intercalated day 3/4,[1] at mid-night; it should have occurred early morning, so that again there is a discrepancy of $\frac{1}{4}$ day. A year later, one occurred in the 33rd year on the fourth intercalated day, in the morning, which in fact was in agreement with the observation before it. Three years later, one occurred in the 36th year on the fourth intercalated day, evening; it should have occurred at mid-night, so that, again, there is a discrepancy of $\frac{1}{4}$ day only.

H196

Observations 3.7 – 3.20, Calculations 3.6 – 3.7

Next he sets out as well the vernal equinoxes that were observed with the same accuracy. One occurred in the 32nd year of the third Calippic cycle, Mechir 27, early morning. And the armillary sphere in Alexandria, he says, was illuminated equally from each direction near the 5th hour; so that in fact the same equinox, observed differently, differed by 5 hours, most nearly. And he says that the subsequent observations up to the 37th year agree with the $\frac{1}{4}$ day surplus. Eleven years later, in the 43rd year, Mechir 29/30, after mid-night, he says the vernal equinox occurred, which was, in fact, in agreement with the observation in the 32nd year and concurs, he says, again with the observations in the next years,

1 *intercalated day 3/4*: This is a short-hand expression: *τῇ τρίτῃ τῶν ἐπαγομένων εἰς τὴν τετάρτην*; literally, "on the third intercalated day approaching the fourth."

up to the 50th year. For it occurred on Phamenoth 1, about the setting of the sun $1\frac{3}{4}$ days later, most nearly, than that in the 43rd year, which also corresponds with the intervening 7 years.

Therefore, in these observations no significant difference has arisen, although it is possible that, concerning not only the tropic but also the equinoctial observations, some error arises in them even up to $\frac{1}{4}$ day. For even if the position or the division of the instruments deviates from precision by $\frac{1}{3600}$ only of the circle through the poles of the equatorial circle, the sun is corrected for a latitudinal recession of such a size, on the equatorial segments, by $\frac{1}{4}$ degree in longitude, as it moves on the oblique circle. So that the discrepancy also differs up to $\frac{1}{4}$ day, most nearly. And it would err yet more in the case of instruments that are not set up on each occasion and rendered precise by the observations themselves; but when, from a certain starting point, they are established together with their underlying foundations to have a position that is, for the most part, stable, some disturbance to them occurs that goes unnoticed due to the passage of time. One might see an example of this in the bronze armillary spheres we have in the palaestra that seem to have their position in the plane of the equatorial circle. For a distortion of their position, and most especially of the larger, more ancient one, appears to us observers so great that sometimes even twice on the same equinoxes their concave surfaces undergo a change of illumination. H197

But, as a matter of fact, Hipparchus himself does not think that any of these things happens to be a credible basis for suspecting that the annual periods are unequal. Calculating from some eclipses of the moon, he says that he finds that the anomaly of the annual periods, when investigated relative to their mean time, does not contain a greater difference than $\frac{3}{4}$ of one day. This would be worthy of some notice, if in fact it were the case and were not based on erroneous observations due to the very factors he cites. For he calculates, through lunar eclipses observed near some fixed stars, how much the star called Spica is westwards of the autumnal point in each case. And he thinks that he thereby finds that sometimes it is, at its greatest, removed $6\frac{1}{2}$ degrees in his own time, while at other times, at its least, $5\frac{1}{4}$ degrees. He concludes from this that, since it is not possible for Spica to move so much in so short a time, it is probable that the sun, from which Hipparchus investigates the positions of the fixed stars, does not make its periodic return in an equal time. But he has failed to notice that, since his calculation cannot be viable at all without assuming the sun's location during the eclipse, in employing to this end for each eclipse the tropics and equinoxes that he himself accurately observed in those years, he himself makes immediately clear that there is no difference beyond the additional $\frac{1}{4}$ day when one compares the annual periods. H198 H199

Observations 3.21 – 3.26, Calculation 3.8

For, to take one example, from the observation of an eclipse he set out in the 32nd year of the third Calippic cycle, he thinks that he finds Spica westwards of the autumnal point by $6\frac{1}{2}$ degrees; while through the observation of an eclipse in the 43rd year of the same cycle, westwards by $5\frac{1}{4}$ degrees. And he similarly sets out for the above calculations the vernal equinoxes accurately observed in these years, in order that through them he get the sun's locations at the middle times of the eclipses, and from these locations those of the moon, and from those of the moon those of the stars. He says that a vernal equinox occurred in the 32nd year, Mechir 27, early morning, and one in the 43rd year, Mechir 29/30, after mid-night, roughly $2\frac{3}{4}$ days later than that which occurred in the 32nd year. These are as many days as $\frac{1}{4}$ day alone, added for each of the intervening eleven years, produces. Accordingly, if the sun has made its periodic return to the equinoxes in question in neither a greater nor a lesser time than the $\frac{1}{4}$ day surplus, and if it is not possible for Spica to have moved $1\frac{1}{4}$ degrees in so few years, how is it not absurd to employ calculations based on the principles in question to discredit the ⟨principles⟩ themselves that establish them? And how is it not absurd to affix the cause of the impossibility of so great a motion of Spica to nothing else but only to the equinoxes in question, although there are many more things that are capable of producing so great an error; as if the equinoxes were at the same time accurately and inaccurately observed? For it would seem more practicable either that the distances involved in the eclipses themselves of the moon to the nearest stars were conjectured in a rather rough way, or else that the calculations, either of the moon's parallaxes (as part of the inquiry into its apparent locations) or of the sun's motion from the equinoxes to the mean times of the eclipses, were either incorrectly or inaccurately obtained.

H200

But I think that Hipparchus himself has recognized that nothing in such considerations is credible enough to attach some second anomalistic motion to the sun. But solely out of a love for the truth, I think, he wanted not to be silent about any thing that can in any way whatever lead some people into suspicion. Accordingly, he himself employed hypotheses of the sun and moon on the assumption that the sun has a single, self-same anomalistic motion: namely, motion whereby it returns to the same point in an annual period taken relative to the tropics and equinoxes. And, in assuming that the periods of the sun that we are proposing are of equal times, we do not in any way observe that the appearances relating to the eclipses differ significantly from what is calculated according to the hypotheses we are proposing. Such a difference would, in fact, turn out to be quite perceptible, if it were even of only one degree or two equatorial hours, most nearly, unless the correction to the inequality of the annual period is included.

H201

Now, on the basis of all these considerations and those from which we ourselves ascertain the times of the periodic returns through the passages of the sun we have successively observed, neither do we find that the annual period is unequal (if it is observed relative to some single thing, and not at one time to the tropic and equinoctial points, while at another time relative to the fixed stars), nor do we find another more appropriate periodic return than that which carries the sun from some tropic point, equinoctial point, or even some other point of the mid-zodiac circle, to the same point again. And, in general, we think it proper to demonstrate the appearances through hypotheses that are as simple as possible, so far as nothing significant in the observations appears to conflict with an assumption of this kind.

Therefore, that the annual period, observed relative to the tropics and the equinoxes, is less than 365 days with the addition of $\frac{1}{4}$ day has become clear to us through those considerations on the basis of which Hipparchus also made his demonstration. But by how much less it is might not be possible to attain with the greatest certainty. The $\frac{1}{4}$ day augmentation remains perceptibly undistinguishable for many years because of the minute quantity of the difference; and, for this reason, the surplus of days that is found, which it is necessary to divide by the years within the interval, can be observed to be the same both in a greater and in a lesser number of years in virtue of the comparison's being made over a rather long period of time. But this sort of periodic return could be obtained with the nearest accuracy the greater the time between the observations one compares is found to be. And this happens not only in this return, but also in all periodic returns. For the error produced by the inadequacy of the observations themselves, even if they are accurately handled, is slight and most nearly the same, so far as perceiving them goes, for both what appears over a long time and what appears over a short time. This error, when divided by fewer years, makes the annual error and the error accumulated from this over a longer time greater, but makes the errors smaller when divided by a greater number of years. H202

Hence it is fitting to believe that it is sufficient if we ourselves, without wittingly neglecting proper examination, attempt to contribute no less than what the time intervening between us and such accurate ancient observations as we possess may add to bringing hypotheses on the annual period into close agreement. And it is also fitting to believe that it is sufficient if we consider assertions about eternity or even a span of time many times greater by far than that covered by the observations to be foreign to a love of learning and a love of truth. H203

As far, then, as regards antiquity, the summer tropics observed by the school of Meton and Euktemon and those by the school of Aristarchus after them, ought to be cited for comparison with what has occurred in our time.

Since the observations of the tropics are in general hard to distinguish,[1] and, in addition, those that have been handed down by those men were taken in a rather rough way, as it seems to appear to Hipparchus as well, we rejected them. But for the present comparison we have made use of observations of the equinoxes and, on account of their accuracy, those that Hipparchus especially noted he had taken with the greatest certainty. And we have made use of those that we ourselves have most indisputably observed through the instruments that we indicated for such purposes at the beginning of the work {H64.12–67.16}. From these observations we find that in 300 years, most H204 nearly, the tropics and equinoxes occur one day earlier than 365 days plus $\frac{1}{4}$ day.

Observations 3.27 – 3.28, Calculation 3.9

For in the 32nd year of the 3rd Calippic cycle, Hipparchus noted most clearly that the autumnal equinox was observed as accurately as possible and says that he has calculated that it occurred on intercalated day 3/4, at midnight; and it is the 178th year from the death of Alexander. 285 years later, in the 3rd year of Antoninus, which is the 463rd year from the death of Alexander, we observed with the greatest certainty, again, that the autumnal equinox occurred on Athur 9, one hour later, most nearly, than the rising of the sun. Therefore, the periodic return for 285 whole Egyptian years, that is to say, those of 365 days, added $70\frac{3}{10}$ days in all, most nearly, instead of the $71\frac{1}{4}$ days that according to the $\frac{1}{4}$ day surplus correspond with the preceding years. So that the periodic return occurred earlier than the $\frac{1}{4}$ day surplus by one day lacking $\frac{1}{20}$ day, most nearly.[2]

Obervations 3.27 – 3.28, Calculations 3.10 – 3.11

In like manner, Hipparchus again says that the vernal equinox, in the above 32nd year of the 3rd Calippic cycle, was most accurately observed to have occurred on Mechir 27, early morning; and it is the 178th year from the death of Alexander. H205 But we find that the vernal equinox, likewise 285 years later, in the 463rd year from the death of Alexander, occurred on Pachon 7, one hour, most nearly, after noon. So that this period too added the same $70\frac{3}{10}$ days, most nearly, instead of the $71\frac{1}{4}$ days that, according to the $\frac{1}{4}$ day surplus, correspond to 285 years. Therefore, in this case too the periodic return of the vernal equinox occurred earlier than the $\frac{1}{4}$ day surplus by one day lacking $\frac{1}{20}$ day. So that, since 300 years have to 285 years the same ratio that 1 day has to 1 day lacking $\frac{1}{20}$ day, it is concluded that in 300

[1] *the observations of the tropics are in general hard to distinguish*: Whether the sun is observed on the horizon or by means of the gnomon's noon shadow—both methods are noted on page 7 of the Preliminaries to *The Almagest*—its change from day to day is very small as it passes through the tropic (solstice) position. The exact day of the tropic is therefore difficult to determine by direct observation.

[2] Thus 300 years equal $(300 \times 365\frac{1}{4} - 1)$ days, so that one year equals $(365\frac{1}{4} - \frac{1}{300})$ days or $365^d\ 5^h\ 55\frac{1}{5}^m$.

years, most nearly, the periodic return of the sun, which occurs relative to the equinoctial points, is also earlier than the $\frac{1}{4}$ day surplus by one day.

Observations 3.29 – 3.30 Calculations 3.12 – 3.13

Even if, for antiquities' sake, we compare the rather roughly recorded summer tropic observed by the school of Meton and Euktemon with that calculated as indisputably as possible by us, we will find this same result. For that is recorded as having occurred when Apseudes was archon in Athens, by Egyptian reckoning Phamenoth 21, early morning, while we calculated with certainty that the summer tropic, in the above 463rd year from the death of Alexander, occurred on Mesore 11/12, two hours, nearly, after mid-night. And there are 152 years from the summer tropic recorded in the time of Apseudes up till that observed by the school of Aristarchus in the 50th year of the first Calippic cycle, just as Hipparchus says, while there are 419 years from the above 50th year, which was during the 44th year from the death of Alexander, up to the 463rd year, that of our observation. Therefore, in the intervening 571 years of the whole interval, if the summer tropic that was observed by the school of Euktemon were to have occurred near the beginning of Phamenoth 21, there are added to whole Egyptian years $140\frac{5}{6}$ days, most nearly, instead of $142\frac{3}{4}$ of those that correspond to the 571 years according to the $\frac{1}{4}$ day surplus; so that the above periodic return occurred earlier than the $\frac{1}{4}$ day surplus by two days lacking $\frac{1}{12}$ day. Therefore, in this way too it has become clear that in 600 whole years the annual period anticipates the $\frac{1}{4}$ day surplus by two full days, most nearly. And through other, more numerous observations, we find that this same thing results and see that Hipparchus is frequently in agreement with it. For in his *On the Magnitude of the Annual Period*, after he compares the summer tropic that was observed by Aristarchus in the 50th year at the end of the first Calippic cycle with that taken accurately by himself, again, in the 43rd year at the end of the third Calippic cycle, he says:

H206

H207

> It is clear, therefore, that in 145 years the tropic occurs earlier than the $\frac{1}{4}$ day surplus by half the time of a day and a night taken together.

And again, in his *On Intercalated Months and Days* as well, after he first states that, according to the school of Meton and Euktemon, the annual period contains $365\frac{1}{4}$ days and $\frac{1}{76}$ day, while according to Calippus $365\frac{1}{4}$ days only, he adds *verbatim*:

> We find that as many whole months are contained in 19 years as those men too found, and, furthermore, that the year also adds less than $\frac{1}{4}$ by $\frac{1}{300}$, roughly, of one day. So that in 300 years it falls short 5 days as compared with Meton, and 1 day as compared with Calippus.

And while roughly summarizing his own thoughts in the course of recording his own treatises, he thus says:

> I have also written about the annual period in one book, in which I prove that it is the solar year; and this period is the time in which the sun arrives from a tropic to the same tropic or from an equinox to the same equinox. It contains 365 days and less than $\frac{1}{4}$ by $\frac{1}{300}$, most nearly, of one day and night; and not according as the mathematicians believe that $\frac{1}{4}$ day itself is intercalated upon the stated quantity of days.

H208

I think, then, that it has become clear that the appearances up to the present relating to the magnitude of the annual period concur, in light of the agreement between present appearances and those of the past, with the quantity previously stated for the periodic return of the tropic and equinoctial points. Since these things are so, if we divide one day by 300 years, there corresponds to each year 0;0,12 of one day. If we subtract this from the 365;15 days (with the $\frac{1}{4}$ day surplus), we will have the sought-for annual period as 365;14,48 days.[1] Such, then, would be the number of days ⟨in the sun's annual period⟩, most nearly, which we have obtained as best we can under the present circumstances.

Calculation 3.14

In order to pursue the inquiry into the individual passages of the sun and the rest of the planets, which the construction of a particular table naturally renders easy and, as it were, open to view, we think that it ought to be an end and goal for a mathematician to show that all the appearances in the heavens are produced through uniform, circular motions.[2] And we think that what is appropriate to and most follows from this end is a table that distinguishes the particular uniform motions from the anomalistic motion[3] that seems to result through the assumption of circles. The table again demonstrates the apparent passages ⟨of the sun and the rest of the planets⟩ through the mixing and combination of both these motions. In order, then, that this kind of table be found

---

[1] The figure 365;14,48 days is equivalent to the $365^{d}\,5^{h}\,55\frac{1}{5}^{m}$ obtained in the preceding calculation. Today the accepted figure for the annual period, expressed sexagesimally, is 365;14,32 days.

[2] *all the appearances in the heavens are produced through uniform, circular motions*: Here Ptolemy states the fundamental axiom of his astronomy. At H216 he reasserts and elaborates this axiom. What he means by it, in detail, will become clearer in the course of Chapter 3.

[3] *anomalistic motion*: The Greek word translated "uniform" is ὁμαλός. It may also be rendered as "regular," "mean," or "average." Its privative form, ἀνομαλός (irregular), is usually translated with the technical term "anomalistic." Since it is a fundamental axiom of Ptolemy's account that the motions in the heavens are uniform and circular, any observed irregularity (anomaly) must be accounted for.

more useful for us and ready at hand for the demonstrations proper, we will set out the particular uniform motions of the sun in the following way.

Calculations 3.15 – 3.19

Now, since the single periodic return has been demonstrated as 365;14,48 days {H208.11–12}, if we divide this into the 360 degrees of one circle, we will have as the daily mean motion of the sun ⟨on the mid-zodiac circle⟩ 0;59,8,17,13,12,31 degrees, most nearly; for it will suffice to make fractional divisions of them up to so many 60ths. H209

Again, taking $\frac{1}{24}$ of the daily motion, we will have as the hourly motion 0;2,27,50,43,3,1 degrees, most nearly.

Similarly, by multiplying the daily motion by the 30 days of one month, we will have as monthly mean motion 29;34,8,36,36,15,30 degrees; while by multiplying it by the 365 days of one Egyptian year, we will have as annual mean motion 359;45,24,45,21,8,35 degrees.

Again, by multiplying the annual motion by 18 Egyptian years, owing to the common measure of the table that will become clear, and subtracting whole cycles, we will have as surplus of an 18-year period 355;37,25,36,20,34,30 degrees.

Accordingly, we arranged three tables of the uniform motion of the sun, each in 45 rows again and in 2 columns. The first table will contain the mean motions of 18-year periods. The second will contain, first, the annual mean motions and under them the hourly motions. The third will contain, first, the monthly motions and the daily mean motions below. The time values are arranged in the first columns and the degrees, with the successive additions proper to each, are set alongside them in the second columns. And the tables are as follows.

2. Table of the Uniform Motion of the Sun.

The distance from the apogee of the sun (5;30 degrees in Gemini) to the mean epoch of the sun (in the first year of Nabonassar, 0;45 in Pisces) is 265;15 degrees.[1] H210

18-year periods	degrees	1st	2nd	3rd	4th	5th	6th
18	355	37	25	36	20	34	30
36	351	14	51	12	41	9	0
54	346	52	16	49	1	43	30
72	342	29	42	25	22	18	0
90	338	7	8	1	42	52	30
108	333	44	33	38	3	27	0

[1] This one sentence conflates two related sentences {H210.2} and {H213.46–48}.

18-year periods	degrees	1st	2nd	3rd	4th	5th	6th
126	329	21	59	14	24	1	30
144	324	59	24	50	44	36	0
162	320	36	50	27	5	10	30
180	316	14	16	3	25	45	0
198	311	51	41	39	46	19	30
216	307	29	7	16	6	54	0
234	303	6	32	52	27	28	30
252	298	43	58	28	48	3	0
270	294	21	24	5	8	37	30
288	289	58	49	41	29	12	0
306	285	36	15	17	49	46	30
324	281	13	40	54	10	21	0
342	276	51	6	30	51	55	30
360	272	28	32	6	12	30	0
378	268	5	57	43	30	4	30
396	263	43	23	19	32	39	0
414	259	20	48	55	53	13	30
432	254	58	14	32	13	48	0
450	250	35	40	8	34	22	30
468	246	13	5	44	54	57	0
486	241	50	31	21	15	31	30
504	237	27	56	57	36	6	0
522	233	5	22	33	56	40	30
540	228	42	48	10	17	15	0
558	224	20	13	46	37	49	30
576	219	57	39	22	58	24	0
594	215	35	4	59	18	58	30
612	211	12	30	35	39	33	0
630	206	49	56	12	0	7	30
648	202	27	21	48	20	42	0
666	198	4	47	24	41	16	30
684	193	42	13	1	1	51	0
702	189	19	38	37	22	25	30
720	184	57	4	13	43	0	0
738	180	34	29	50	3	34	30

18-year periods	degrees	1st	2nd	3rd	4th	5th	6th
756	176	11	55	26	24	9	0
774	171	49	21	2	44	43	30
792	167	26	46	39	5	18	0
810	163	4	12	15	25	52	30
828*	158	41	37	51	46	27	0
846*	154	19	3	28	7	1	30
864*	149	56	29	4	27	36	0
882*	145	33	54	40	48	10	30
900*	141	11	20	17	8	45	0

*Ptolemy's table extended only to 810 years. Additional entries have here been supplied to facilitate modern use.

simple years	sun's degrees	1st	2nd	3rd	4th	5th	6th
1	359	45	24	45	21	8	35
2	359	30	49	30	42	17	10
3	359	16	14	16	3	25	45
4	359	1	39	1	24	34	20
5	358	47	3	46	45	42	55
6	358	32	28	32	6	51	30
7	358	17	53	17	28	0	5
8	358	3	18	2	49	8	40
9	357	48	42	48	10	17	15
10	357	34	7	33	31	25	50
11	357	19	32	18	52	34	25
12	357	4	57	4	13	43	0
13	356	50	21	49	34	51	35
14	356	35	46	34	56	0	10
15	356	21	11	20	17	8	45
16	356	6	36	5	38	17	20
17	355	52	0	50	59	25	55
18	355	37	25	36	20	34	30

hours	degrees	1st	2nd	3rd	4th	5th	6th
1	0	2	27	50	43	3	1
2	0	4	55	41	26	6	2
3	0	7	23	32	9	9	3

hours	degrees	1st	2nd	3rd	4th	5th	6th
4	0	9	51	22	52	12	5
5	0	12	19	13	35	15	6
6	0	14	47	4	18	18	7
7	0	17	14	55	1	21	9
8	0	19	42	45	44	24	10
9	0	22	10	36	27	27	11
10	0	24	38	27	10	30	12
11	0	27	6	17	53	33	14
12	0	29	34	8	36	36	15
13	0	32	1	59	19	39	16
14	0	34	29	50	2	42	18
15	0	36	57	40	45	45	19
16	0	39	25	31	28	48	20
17	0	41	53	22	11	51	21
18	0	44	21	12	54	54	23
19	0	46	49	3	37	57	24
20	0	49	16	54	21	0	25
21	0	51	44	45	4	3	27
22	0	54	12	35	47	6	28
23	0	56	40	26	30	9	29
24	0	59	8	17	13	12	31

Egyptian months*	sun's degrees	1st	2nd	3rd	4th	5th	6th
30	29	34	8	36	36	15	30
60	59	8	17	13	12	31	0
90	88	42	25	49	48	46	30
120	118	16	34	26	25	2	0
150	147	50	43	3	1	17	30
180	177	24	51	39	37	33	0
210	206	59	0	16	13	48	30
240	236	33	8	52	50	4	0
270	266	7	17	29	26	19	30
300	295	41	26	6	2	35	0
330	325	15	34	42	38	50	30
360	354	49	43	19	15	6	0

*Entries in the column labeled "Egyptian months" are actually in days, each 30-day period making up an Egyptian month.

days	degrees	1st	2nd	3rd	4th	5th	6th
1	0	59	8	17	13	12	31
2	1	58	16	34	26	25	2
3	2	57	24	51	39	37	33
4	3	56	33	8	52	50	4
5	4	55	41	26	6	2	35
6	5	54	49	43	19	15	6
7	6	53	58	0	32	27	37
8	7	53	6	17	45	40	8
9	8	52	14	34	58	52	39
10	9	51	22	52	12	5	10
11	10	50	31	9	25	17	41
12	11	49	39	26	38	30	12
13	12	48	47	43	51	42	43
14	13	47	56	1	4	55	14
15	14	47	4	18	18	7	45
16	15	46	12	35	31	20	16
17	16	45	20	52	44	32	47
18	17	44	29	9	57	45	18
19	18	43	37	27	10	57	49
20	19	42	45	44	24	10	20
21	20	41	54	1	37	22	51
22	21	41	2	18	50	35	22
23	22	40	10	35	3	47	53
24	23	39	18	53	17	0	24
25	24	38	27	10	30	12	55
26	25	37	35	27	43	25	26
27	26	36	43	44	56	37	57
28	27	35	52	2	9	50	28
29	28	35	0	19	23	2	59
30	29	34	8	36	36	15	30

3. On the Hypotheses as to Uniform Circular Motion. H215

H216.3 Since the next task is to explain the apparent anomalistic motion of the sun, it is necessary to premise in general that the motions of the planets eastwards[1] in the heavens, just as the movement of all the stars westwards, are all uniform and circular by nature. That is to say, the straight lines, conceived of as revolving the stars or even their circles, intercept equal angles in equal times at the centers of each revolution in all cases without exception. But the anomalistic motions that appear with them are brought about due to the positions and orderings of the circles themselves in their spheres, through which they produce their motions; and none of them, for all the suspected disorderliness of their appearances, in reality turns out by nature to be foreign to eternity.

The cause of their anomalistic appearance may arise on roughly two primary, simple hypotheses. For their motion is observed relative to the circle that is conceived of as being concentric with the universe and in the plane of the mid-zodiac circle, such that our sight does not appreciably differ from its center. Hence one must suppose that they make motions that are uniform either on circles non-concentric with the universe, or else on circles that are concentric with it but not upon them simply but, rather, upon other circles carried by them and called "epicycles." For on each of these hypotheses, the fact that in equal times they seem to traverse, to our eyes, unequal arcs of the mid-zodiac circle that is concentric with the universe will appear possible.

H217

Proposition 3.1

For, on the eccentrical hypothesis, if we conceive of the eccentric circle, upon which the star moves uniformly, *ABCD* around center *E* and diameter *AED*, and the point *F* upon it is our sight, so that also *A* is the point of farthest apogee, while *D* of nearest perigee;[2] and if, after intercepting equal arcs *AB* and *DC*, we join *BE*, *BF*, *CE*, and *CF*, it will be immediately clear that the star, moving along each of the arcs *AB* and *CD* in an equal time, will seem to have traversed unequal arcs of the circle described around center *F*.

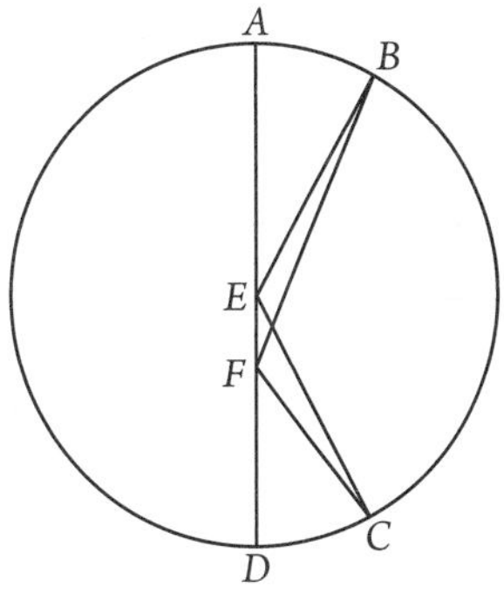

For, since angle *BEA* is equal to angle *CED* [Euc. 3.26], angle *BFA* is smaller than each of them [Euc. 1.16], but angle *CFD* is greater.

---

1 *the motions of the planets eastwards*: Looking down from above the north pole of the sun's oblique sphere, "eastwards" would be the counter-clockwise direction; "westwards" denotes the clockwise direction around the poles of the celestial sphere, the direction of the diurnal motion. Recall that for Ptolemy, "the planets" include the sun and the moon.

2 *farthest apogee ... nearest perigee*: Ptolemy here uses the superlative forms *ἀπογειότατον*, "farthest apogee," and *περιγειότατον*, "nearest perigee," that is, greatest and least distances from the earth. The simple forms "apogee" and "perigee" are more normal.

Proposition 3.2

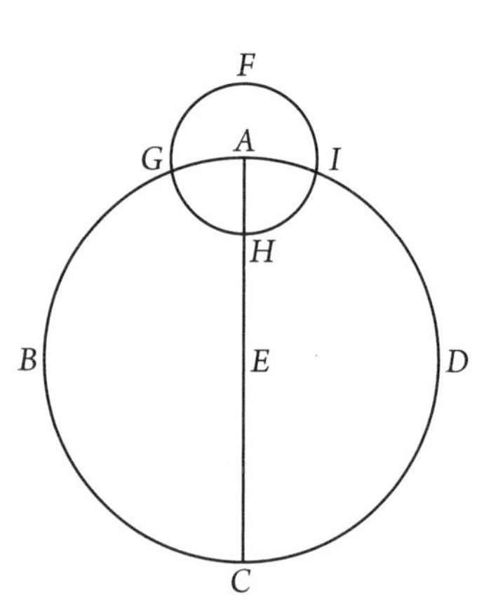

H218

And, on the epicyclical hypothesis, if we conceive of the circle *ABCD* concentric with the mid-zodiac circle around center *E* and diameter *AEC*, and the epicycle *FGHI* around center *A* (upon which the star moves) being carried upon it, it will be immediately clear in this way also that, when the epicycle uniformly traverses circle *ABCD*, as, for example, from *A* to *B*, and the star the epicycle, when the star is at *F* and *H*, it will appear indistinguishably at the center of the epicycle, *A*.

But when it is at other points, this is no longer the case; but when it is at, let us say, *G*, it will seem to have made a motion greater than uniform motion by arc *AG*, but when at *I* a smaller motion than uniform motion, in the same way, by arc *AI*.

Propositions 3.3 – 3.4

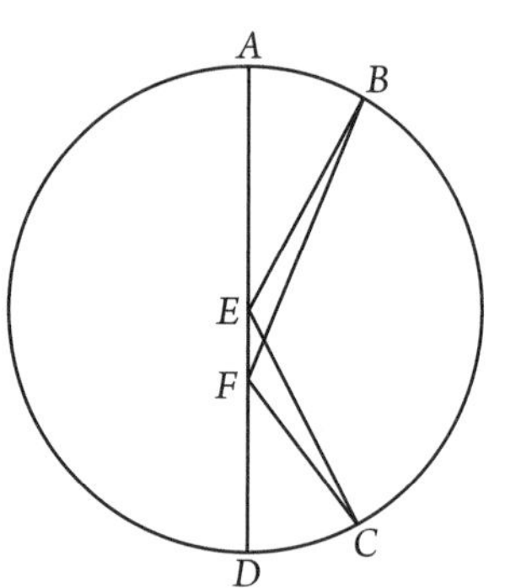

Figure for Prop. 3.1

H219

Upon this sort of eccentrical hypothesis,[1] then, it always results that least motion holds at the point of farthest apogee, while greatest motion holds at the point of nearest perigee,[2] since ⟨in the figure for Prop. 3.1⟩ angle *AFB* is always less than angle *DFC*. On the epicyclical hypothesis both can occur. For ⟨in the figure for Prop. 3.2⟩ as the epicycle makes its motion eastwards in the heavens, as, for example, from *A* to *B*, if the star on its epicycle so makes its motion that the motion from apogee is again produced eastwards, that is to say from *F* toward *G*, it will result that greatest passage occurs at apogee, because at that time both the epicycle and the star move in the same direction. But if the motion from the apogee of the star occurs westwards on the epicycle, that is to say from *F* toward *I*, least passage will, contrariwise, be produced at apogee, because at that time the star moves in the direction opposite to the motion of the epicycle.

Since these things are so, next it is necessary to premise the following matters as well. For stars that make double anomalistic motions, it is possible

---

[1] *this sort of eccentrical hypothesis*: In Book XII Ptolemy will introduce a different kind of eccentrical hypothesis.

[2] *least motion ... greatest motion*: The planet's least *apparent* motion occurs when it is furthest from the earth, and its greatest *apparent* motion occurs when it is closest to the earth.

to combine both these hypotheses, as we will prove in our remarks concerning them {IX.2}, while for stars that are subject to one and the same anomalistic motion, just one of the above hypotheses will suffice.

And all the appearances will be produced indistinguishably on each hypothesis, when the same ratios are maintained in both, namely: when the radius of the epicycle has to the radius of the circle that carries it[1] (on the epicyclical hypothesis) the same ratio that the line between the center of sight and the center of the eccentric circle[2] has to the radius of the eccentric circle (on the eccentrical hypothesis); and furthermore, when the epicycle, again moving eastwards, traverses the circle concentric with sight and the star traverses the epicycle (on the assumption, however, that the motion ⟨of the star on the epicycle⟩ at apogee is westwards) in the same amount of time in which the star, moving eastwards, travels with the same speed[3] over the eccentric circle that remains stationary.

Proposition 3.5

H220

The fact that, when these things are assumed in this way, the same appearances will result on each of the hypotheses, we will briefly work out both through their ratios and afterwards {III.4} through the numeric values treated in them for the anomalistic motion of the sun.

Now, I say first that, on each hypothesis, the greatest difference of ⟨the arc of⟩ uniform motion as compared with ⟨the arc of⟩ apparent anomalistic motion, according to which the mean passage of the stars[4] is conceived, occurs when the apparent distance from apogee cuts off a quadrant, and that the time from farthest apogee to the stated mean passage is greater than the time from mean passage to nearest perigee. Hence it results, on the hypothesis of eccentric circles always, and on the hypothesis of epicycles when the motions ⟨of the stars on their epicycles⟩ occur from their apogees westwards, that the time from least motion to mean motion is greater than the time from mean motion to greatest motion, because least passage is produced at apogee on each hypothesis. But on the hypothesis that produces

H221

1 *the circle that carries it*: the circle that carries the epicycle is traditionally called the *deferent*.

2 *the line between the center of sight and the center of the eccentric circle*: The distance represented by this line is called the *eccentricity*.

3 *same speed*: The same angular speed, not the same linear speed. This stipulation has the effect of making the line from the epicycle's center to the planet remain always parallel to itself (and to the line of apsides) as the epicycle is carried around the concentric circle.

4 *the mean passage*: The occasion on which the apparent movement of the planet eastward through the mid-zodiac circle is equal to its mean, or average, movement, as determined in Book III, Chapter 2. Ptolemy takes it as evident that mean passage occurs at the positions of greatest anomalistic difference; and proves below that each of these positions is an apparent quadrant's distance from apogee.

the revolutions of stars on their epicycles from apogee eastwards, the time from greatest motion to mean motion is, contrariwise, greater than the time from mean motion to least motion, because the greatest passage is produced at apogee in this case.

Proposition 3.5, A.1

Now, first let eccentric circle *ABCD* of the star be around center *E* and diameter *AEC*, upon which let there be taken the center of the zodiac circle, that is to say the visual center, and let it be *F*.

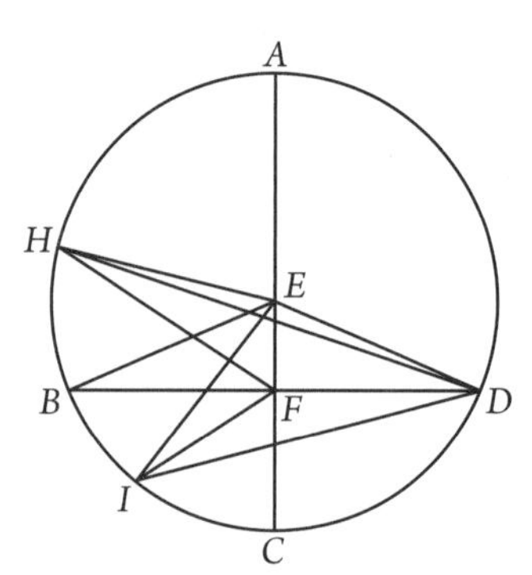

And when line *BFD* has been produced through *F* at right angles to *AEC*, let the star be assumed on points *B* and *D* in order that the apparent distance clearly be removed from apogee *A* by a quadrant on each side.

It must be shown that at points *B* and *D* the greatest difference of ⟨the arc of⟩ uniform motion compared with ⟨the arc of⟩ anomalistic motion arises.

For let *EB* and *ED* be joined.

The fact, then, that the arc of anomalistic difference has to the whole circle the ratio that angle *EBF* has to four right angles[1] is immediately clear, since angle *AEB* subtends the arc of uniform motion, angle *AFB* the arc of apparent anomalistic motion, and angle *EBF* is their difference [Euc. 1.32].

H222

Now, I say that another angle greater than either angle *EBF* and angle *EDF* will not be constructed on the circumference of circle *ABCD* upon the straight line *EF*.

For let angles *EHF* and *EIF* be constructed at points *H* and *I*, and let *HD* and *ID* be joined.

Accordingly, since in every triangle the greater side subtends the greater angle [Euc. 1.19]
and *HF* is greater than *FD* [Euc. 3.7, Euc. 3.3],
angle *HDF* also will be greater than angle *DHF*.

And angle *EDH* is equal to angle *EHD* [Euc. 1.5], since *EH* is equal to *ED*; and therefore the whole angle *EDF*, that is to say angle *EBD*, is greater than angle *EHF*.

Again, since *DF* is greater than *IF*,
angle *FID* is also greater than angle *FDI*.

But angle *EID* is equal to the whole angle *EDI*,
since *EI* again is equal to *ED*; and therefore the remaining angle *EDF*, that is to say angle *EBF*, is greater than angle *EIF*.

---

[1] *the ratio that angle EBF has to four right angles*: Ptolemy is here speaking of angle *EBF* as if it were subtending an arc of a circle around *B* as center.

H223 Therefore, it is not possible to construct other greater angles, in the manner we have stated, than those at points *B* and *D*.

And it is proved at the same time that arc *AB*, which contains the time from least motion to mean motion, is greater than arc *BC*, which contains the time from mean motion to greatest motion, by two arcs that contain the anomalistic difference; *(Proposition 3.5, A.2)*

since angle *AEB* is greater than a right angle, that is to say than angle *EFB*, by angle *EBF*,

while angle *BEC* is smaller ⟨than a right angle⟩ by the same angle.

Again, in order to show that the same thing results on the other hypothesis also, let circle *ABC*, concentric with the universe, be around center *D* and diameter *ADB*, and epicycle *EFG* being carried upon it in the same plane, be around center *A*. And let the star be assumed at *G*, when it appears to be a quadrant removed from the point at apogee,[1] and let *AG* and *DGC* be joined. *(Proposition 3.5, A.3)*

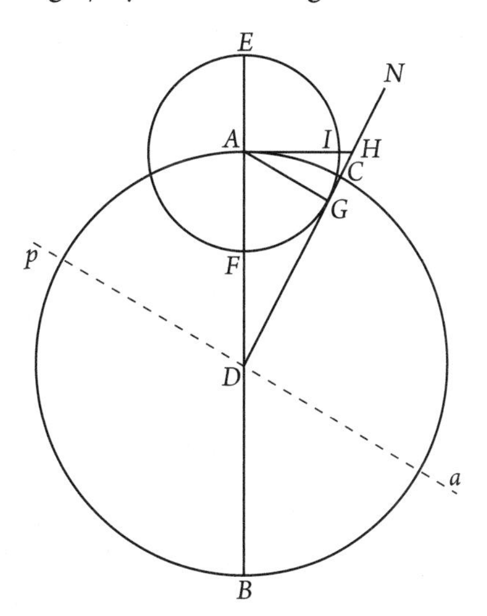

I say that *DGC* is tangent to the epicycle; for then the greatest difference of uniform motion compared with anomalistic motion arises.

H224 For the ⟨arc of⟩ uniform motion ⟨of the star on the epicycle⟩ from apogee is contained by angle *EAG*,[2]

for the star traverses the epicycle and the epicycle the circle *ABC* with equal speed;

but the difference of ⟨the arc of⟩ uniform motion compared with ⟨the arc of⟩ apparent motion is contained by angle *ADG*.

---

1 *apogee*: Ptolemy's diagram did not include any indication of apogee or perigee; however, the line of apsides must be parallel to *AG*; see footnote 3 on "same speed," page 97 above. For convenience, we have indicated the relative position of the line of apsides in this and subsequent diagrams. The apogee is in the direction of *a*, and perigee in the direction of *p*. The star's apparent angular removal from apogee, angle *GDa*, is a right angle.

2 *angle EAG*: Note that Ptolemy here assumes that the star's motion on the epicycle is westwards, as he stipulated in describing the epicyclic form of hypothesis {H220}. This is represented as clockwise motion on the epicycle.

Hence it is clear that the difference between angle *EAG* and angle *ADG*, that is to say angle *AGD*, contains the apparent distance of the star from apogee.[1]

So that, since this distance is assumed to be a quadrant, angle *AGD* will also be right,
and, for this reason, the straight line *DGC* also will be tangent to epicycle *EFG* [Euc. 3.16. porism].

Therefore, arc *AC* between center *A* and the tangent is the greatest anomalistic difference.

Proposition 3.5, A.4

And, by the same considerations, arc *EG*, which contains, according to the motion that in this case is assumed upon the epicycle, the time from least motion to mean motion, is greater than arc *GF*, which contains the time from mean motion to greatest motion, by twice the arc *AC*.

H225

For if we produce *DGH* and draw *AIH* at right angles to *EF*,
angle *IAG* becomes equal to angle *ADC* and arc *IG* similar to arc *AC*,
while arc *EIG* is greater but arc *FG* smaller than one quadrant by this arc:
which it was necessary to prove.

Proposition 3.5, B.1

The fact that, in the case of particular motions also, on each of the hypotheses in equal times all the same things result in terms of uniform and apparent motions and, furthermore, their differences (that is to say the anomalistic difference), one might best understand from the following.

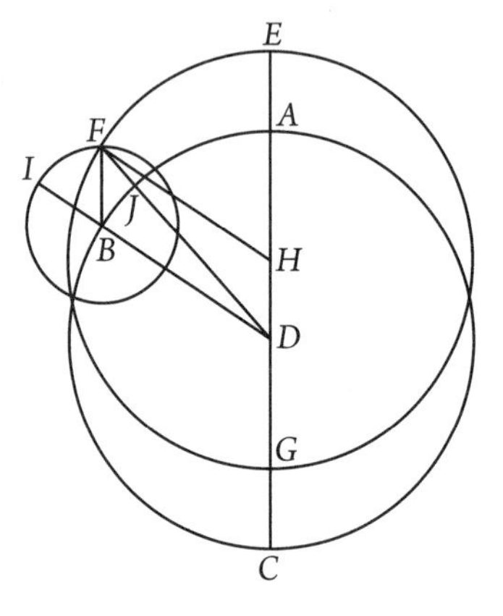

For let circle *ABC*, concentric with the mid-zodiac circle, be around center *D*, and let circle *EFG*, eccentric yet equal to the concentric circle *ABC*, be around center *H*, and let *EAHD* be the common diameter of both, through centers *D* and *H* and apogee *E*. And, when random arc *AB* is intercepted upon the concentric circle, let epicycle *IF* be described with center *B* and distance *DH*, and let IBD be joined.

---

[1] *angle AGD contains the apparent distance of the star from apogee*: Note that this statement is true for *any* location of *G* on circumference of the epicycle. It follows that a line drawn from *D* to apogee *a* will always be parallel to *AG* and in the same direction.

H226 I say that the star will be carried by each of the motions towards the intersection, *F*, of the eccentric circle and the epicycle absolutely in the same time. That is to say, the three arcs, *EF* of the eccentric circle, *AB* of the concentric circle, and *IF* of the epicycle, will be similar to one another, while the difference of ⟨the arc of⟩ uniform motion as compared with ⟨the arc of⟩ anomalistic motion and the apparent passage of the star will turn out to be similar and the same on each of the hypotheses.

For let *FH* and *BF* be joined, and, furthermore, *DF*.

Since in quadrilateral *BDHF* the opposite sides are equal respectively, *FH* to *BD* and *BF* to *DH*, quadrilateral *BDFH* will be a parallelogram.

Therefore, the three angles *EHF*, *ADB*, and *FBI* are equal [Euc. 1.29].

So that, since they are at the centers of circles, the arcs subtended by them also are similar to one another: *EF* of the eccentric circle, *AB* of the concentric circle, and *FI* of the epicycle.

Therefore, on both motions, the star will be carried in an equal time towards the same point *F* and will appear to have traversed the same arc *AJ* of the mid-zodiac circle from apogee.

H227 And consequently the anomalistic difference will also be the same on each of the hypotheses, since we showed that this sort of difference is contained, on the eccentrical hypothesis, by angle *DFH*, while on the epicyclical hypothesis by angle *BDF*; and these also are both equal and alternate, because *FH* was shown to be parallel to *BD*.

Proposition 3.5, B.2

It is clear that for all distances as well the same things will follow, when the ratios happen to be similar and equal on each of the hypotheses, since quadrilateral *HDFB* always is a parallelogram and the eccentric circle is always described by the movement of the star on the epicycle.

Proposition 3.5, C.1

The fact that, even if the ratios are only similar but unequal in magnitude the same appearances will again result[1] will become clear in the following way.

For, in like manner, let circle *ABC*, concentric with the universe, be around center *D* and diameter *ADC*, on which the star is at farthest apogee and nearest perigee, and let the epicycle around *B* be separated from apogee *A* by random arc *AB*. And let the star be moved through arc *EF* that is clearly similar to arc *AB*, because the periodic returns of the circles are ⟨given as⟩ isochronous, and let *DBE*, *BF*, and *DF* be joined.

H228

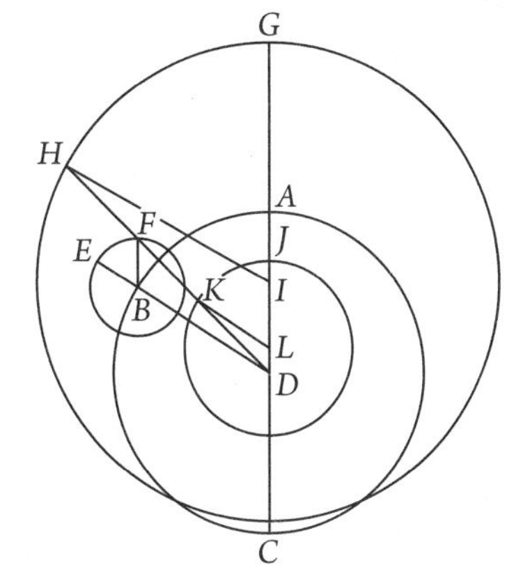

[1] *even if the ratios are only similar...*: Even if the eccentric does not equal the deferent, the same appearances will result so long as the *radius of eccentric* : *radius of deferent* :: *eccentricity* : *radius of epicycle*.

The fact, then, that angles *ADE* and *FBE* are always equal and the fact that the star will appear upon the straight line *DF* are immediately clear on this hypothesis.

I say that, by the eccentrical hypothesis also, if the eccentric circle is either larger or smaller than concentric circle *ABC*, while only the similarity of ratios and the isochronicity of periodic returns are assumed, the star will appear again on the same straight line *DF*.

For let a larger eccentric circle *GH*, as we said, be described around center *I* on *AC*, and a smaller eccentric circle *JK* similarly about center *L*, and, when *DKFH* and *DJAG* are produced, let *HI* and *KL* be joined.

Since *HI* is to *ID* and *KL* is to *LD* as *DB* is to BF {H219.21},

H229

and angle *BFD* is equal to angle *KDL*, because *DA* is parallel to *BF* [Euc. 6.7], the three triangles are equiangular, and the angles subtending the proportional sides, *BDF*, *DHI*, and *DKL*, are equal;

therefore, straight lines *BD*, *HI*, and *KL* are parallel [Euc. 1.28].

So that also, angles *ADB*, *AIH*, and *ALK* are equal [Euc. 1.29].

And since they are at the centers of their circles, the arcs on them, *AB*, *GH*, and *JK*, will be similar.

Therefore, in an equal time not only has the epicycle traversed arc *AB* and the star arc *EF*, but also the star will have traversed arc *GH* and arc *JK* on the eccentric circles and, for this reason, it will always be observed upon the same straight line *DKFH*; that is, being at point *F* on the epicycle, at point *H* on the larger eccentric circle, at point *K* on the smaller eccentric circle, and similarly for all positions.

Proposition 3.5, D.1

It results also that, when the star appears to have intercepted an equal arc from apogee and perigee, the anomalistic difference will also be equal at each position. For, by the eccentrical hypothesis, if we describe eccentric circle *ABCD* around center *E* and diameter *AEC* through apogee *A*, our sight being assumed upon it at point *F*, and, producing random line *BFD* through *F*, we join *EB* and *ED*, the apparent passages will be equal and opposite.

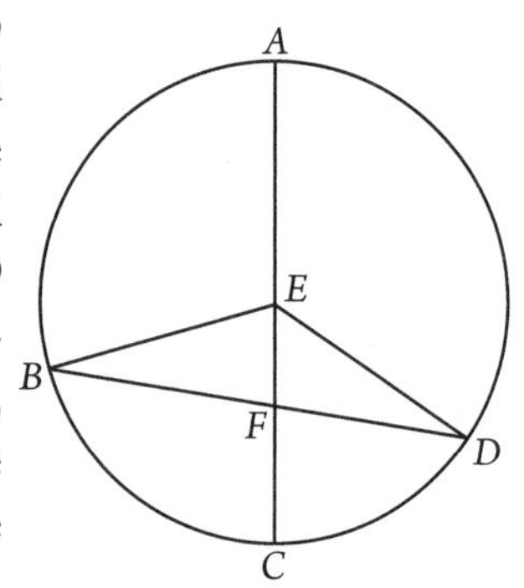

H230

That is to say, angle *AFB* of the ⟨apparent⟩ passage from apogee and angle *CFD* of the ⟨apparent⟩ passage from perigee, and the anomalistic difference will be the same, because *BE* is equal to *ED*, and angle *EBF* is equal to angle *EDF* [Euc. 1.5].

So that the arc of uniform motion from apogee *A* is greater and the arc of uniform motion from perigee *C* is smaller than the apparent arc (that is to say, the arc contained by each of the angles, *AFB* and *CFD*) by the same difference, because angle *AEB* is greater than angle *AFB* and angle *CED* is smaller than angle *CFD* [Euc. 1.32].

Proposition 3.5, D.2

And on the epicyclical hypothesis, if we similarly describe circle *ABC* concentric ⟨with the universe⟩ around center *D* and diameter *ADC* and epicycle *EFG* around center *A*, produce the random line *DGBF*, and join *AF* and

H231

AG, then the arc of anomalistic difference, *AB*, will again be the same, assumed at both positions; that is to say, whether the star is at *F* or at *G*.

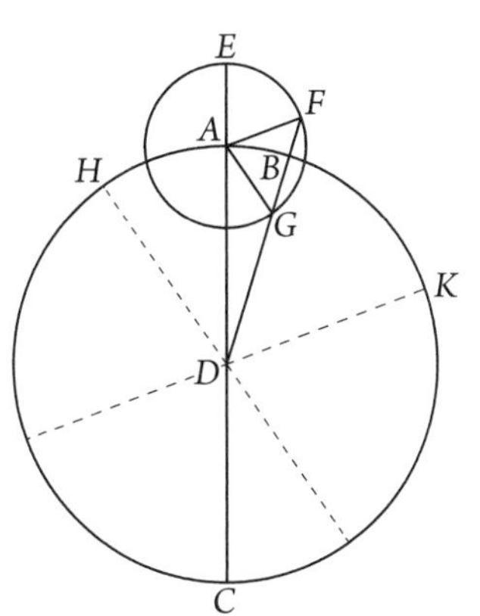

And the star will also appear equally removed from the apogee point of the mid-zodiac circle when it is at *F* and from the perigee point when it is at *G*.

The reasons are: the apparent arc from apogee is contained by angle *DFA*, for it was shown to be the difference between the uniform motion ⟨angle *ADK*⟩ and the anomalistic difference ⟨angle *ADF*⟩;[1]
and the apparent arc from perigee is contained by angle *FGA*, for it is equal to the sum of the uniform motion from perigee ⟨angle *HDA*⟩ and the anomalistic difference ⟨angle *ADG*⟩;[2]
and angle *DFA* is equal to angle *FGA* [Euc. 1.5] because *AF* is equal to *AG*.

Therefore on this hypothesis also it is again concluded that it is by the same difference, that is to say angle *ADG*, that mean motion at apogee is greater than apparent motion, that is to say angle *EAF* is greater than angle

H232

AFD, and mean motion at perigee is smaller than apparent motion which is the same, that is to say angle *GAD* is smaller than angle *AGF* [Euc. 1.32]; which it was proposed to prove.

---

[1] Alternatively, recall that a radius drawn from the epicycle's center to the star will always be parallel to the line of apsides, in the direction towards apogee. Thus when the star is at *F*, the line of apsides coincides with *DK*, where *K* is apogee. Then the star's apparent distance from apogee will be the angle *KDF*, which is equal to angle *DFA*.

[2] As before, when the star is at *G*, the line of apsides coincides with *HD*, where *H* is perigee. Then the star's apparent distance from perigee will be the angle *HDG*, which is equal to angle *FGA*.

4. On the Apparent Anomalistic Motion of the Sun.

Now that these things have been set out beforehand in this way, it must be premised that the apparent anomalistic motion of the sun is single[1] and always makes the time from least motion to mean motion greater than the time from mean motion to greatest motion.[2] For we find that this is also in agreement with the appearances. The apparent anomalistic motion can be produced on each of the proposed hypotheses, albeit on the epicyclical hypothesis when the movement of the sun on its arc at apogee occurs westwards. But it would be more reasonable (εὐλογώτερον) that it be attached to the eccentrical hypothesis, as this hypothesis is simpler and is produced by one, and not by two motions.

Now, the first task is to find the ratio of the eccentricity of the sun's circle, that is to say, what ratio the straight line between the center of the eccentric circle and the visual center of the mid-zodiac circle has to the radius of the eccentric circle, and, furthermore, on precisely what segment of the mid-zodiac circle is the point of farthest apogee of the eccentric circle. Hence these matters have been shown diligently by Hipparchus. For he posited that the time from a vernal equinox to a summer tropic is $94\frac{1}{2}$ days, while the time from a summer tropic to an autumnal equinox is $92\frac{1}{2}$ days.[3] Through these appearances alone he proves that the straight line between the previously stated centers is $\frac{1}{24}$, most nearly, of the radius of the eccentric circle, and its apogee[4] is westwards of the summer tropic by $24\frac{1}{2}$ degrees, most nearly, where the mid-zodiac circle is 360 degrees. We too find that the times of the above quadrants and the above ratios are the same, most nearly, even now. For this reason, the fact that the eccentric circle of the sun always preserves the same position relative to the tropic and equinoctial points[5] also becomes clear to us.

H233

1 *the apparent anomalistic motion of the sun is single*: The sun has only one anomaly, unlike the planets, which have compound anomalies.

2 *the time from least motion to mean motion ... greater than the time from mean motion to greatest motion*: See the Table of Motion of the Apparent Sun in the Preliminaries to Book III, pages 79–80 above. The year charted, 2010, is typical in this respect.

3 On the assumption of regular circular motion around the ecliptic center, the time from the vernal equinox to the summer solstice should equal the time from summer solstice to autumnal equinox and the sum of the two should equal one half year. The discrepancy between this expectation and what is actually observed constitutes the sun's anomaly.

4 The line containing the sun's apogee and perigee is termed the *line of apsides*.

5 *the eccentric circle of the sun always preserves the same position relative to the tropic and equinoctial points*: Ptolemy inferred this since he obtained the same day counts between successive equinoxes and tropics as did Hipparchus. But when modern observations are taken into account, we find that the sun's line of apsides moves eastward at a rate of nearly 2 degrees per century.

In order that this sort of opportunity not be neglected, but also in order that the theorem that has been systematically treated through our numeric values be set out, we will ourselves make a demonstration of the foregoing on an eccentric circle, using the same appearances, namely, as we said: the time from a vernal equinox up to a summer tropic contains $94\frac{1}{2}$ days and the time from a summer tropic to an autumnal equinox contains $92\frac{1}{2}$ days. For, in fact, through the equinoxes and summer tropic we most accurately observed in the 463rd year from the death of Alexander, we find the number of the intervals of days in agreement. Since, as we said {H204.10, 205.2, 206.2}, the autumnal equinox occurred on Athur 9, after the rising of the sun, and the vernal equinox on Pachon 7, after mid-day. So that the interval sums up to $178\frac{1}{4}$ days.

H234

And we find that the summer tropic occurred on Mesore 11/12 after mid-night, so that this interval also, that is to say the interval from the vernal equinox to the summer tropic, totals up to $94\frac{1}{2}$ days. And there is left over for the interval from the summer tropic to the next autumnal equinox the remaining $92\frac{1}{2}$ days, most nearly, of the annual period.

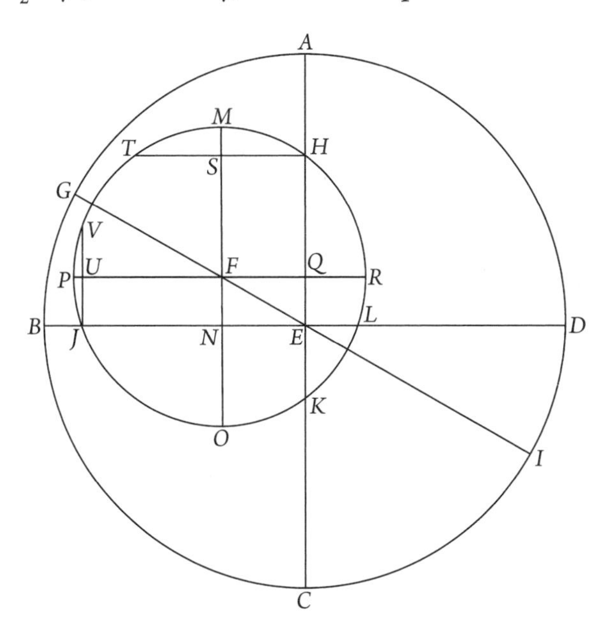

Calculation 3.20

Now, let the mid-zodiac circle *ABCD* be around center *E*, and let there be drawn through in it two diameters *AC* and *BD*, at right angles to one another, through the tropic and equinoctial points, and let it be supposed that *A* is the vernal point, *B* the summer point, and the rest accordingly.[1]

[1] *the vernal point ... the summer point*: The vernal equinox occurs at the beginning of the zodiac sign Aries; the summer solstice occurs at the beginning of Cancer.

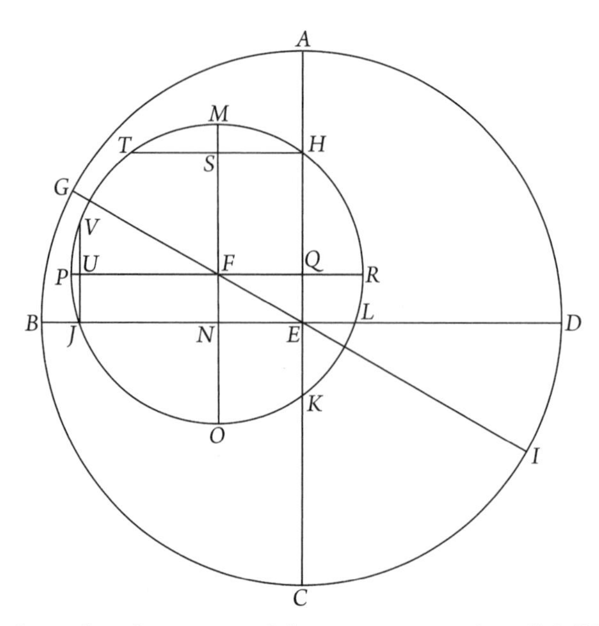

The fact, then, that the center of the eccentric circle will fall between the straight lines *EA* and *EB* is clear; because semicircle *ABC* contains a greater period than half the annual period and, for this reason, intercepts a segment of the eccentric circle greater than a semicircle, and because quadrant *AB* itself contains a greater period and intercepts a greater arc of the eccentric circle compared with quadrant *BC*. H235

Since this is so, let ⟨some⟩ point *F* be supposed as center of the eccentric circle, and let diameter *EFG*, through both centers and the apogee, be drawn through, and with center *F* and with a random distance let the eccentric circle of the sun, *HJKL*, be described. And through *F* let *MNO* be drawn parallel to *AC* and *PQR* to *BD* and, furthermore, let *HST* be drawn perpendicular from *H* to *MNO*, and *JUV* from *J* to *PQR*.

Therefore, since the sun, uniformly traversing circle *HJKL*, travels arc *HJ* in $94\frac{1}{2}$ days and arc *JK* in $92\frac{1}{2}$ days, and in $94\frac{1}{2}$ days it uniformly moves 93;9 degrees, most nearly,[1] where the circle is 360 degrees,
while in $92\frac{1}{2}$ days ⟨it uniformly moves⟩ 91;11 degrees,
segment *HJK* would be 184;20 degrees.

Segments *MH* and *KO* taken together would be the 4;20 degrees that remain over and above the semicircle *MPO*, H236
and *HMT*, the arc double the arc *HM* [Euc. 3.3], would be 4;20 of the same degrees.

---

1 *93;9 degrees, most nearly*: At H203–4 above the length of the year was found to be $365\frac{1}{4} - \frac{1}{300}$ days, or $365\frac{37}{150}$ days. Dividing the time of arc *HJ* ($94\frac{1}{2}$ days) by the length of the year ($365\frac{37}{150}$ days) and multiplying by 360, we have the length of the arc in degrees, 93;8,55.

So that also, the chord subtending it, *HT*, will be 4;32 parts, most nearly, where the diameter of the eccentric circle is 120 parts,
and half of it, *HS*, that is to say, *EN*, will be 2;16 of the same parts.

Again, since the whole segment *HMPJ* is 93;9 degrees,
segment *HM* 2;10 degrees,
and quadrant *MP* 90 degrees,
then the remaining arc *PJ* will be 0;59 degrees,
and the arc double it, *JPV*, will be 1;58 degrees.

So that also, the chord subtending it, *JUV*, will be 2;4 parts where the diameter of the eccentric circle is 120 parts,
while half of it, *JU*, that is to say, *FN*, will be 1;2 parts.

Straight line *EN* also was shown to be 2;16 of the same parts. And since the squares on them when added together make the square on *EF* [Euc. 1.47], *EF* itself in length will be 2;29$\frac{1}{2}$ parts, most nearly, where the radius of the eccentric circle is 60.

Therefore, the radius of the eccentric circle is 24 times, most nearly, the line between its center and the center of the zodiac circle.

H237 Again, since straight line *FN* also is 1;2 parts where *EF* was shown to be 2;29$\frac{1}{2}$ parts, Calculation 3.21
then, accordingly, straight line *FN* will be 49;46 parts, most nearly, where hypotenuse *EF* is 120 parts,
while the arc on it of the circle described around right-angled triangle *EFN* will be 49 degrees, most nearly, where the circle is 360 degrees.

And therefore angle *FEN* will be 49 degrees where two right-angles are 360 degrees,[1] and 24;30 degrees where four right-angles are 360 degrees.

So that, since angle *FEN* is at the center of the zodiac circle, then arc *BG*, by which the point at apogee *G* is westwards of the summer tropic point *B*, is 24;30 degrees.

For the rest, since quadrant *OR* and quadrant *RM* are each 90 degrees,
and arc *OK* itself and arc *HM* are each 2;10 degrees,
while arc *LR* is 0;59 degrees,
then arc *KL* will be 86;51 degrees and arc *LH* 88;49 degrees.

But the sun uniformly traverses 86;51 degrees in 88$\frac{1}{8}$ days and 88;49 degrees in 90$\frac{1}{8}$ days, most nearly.

---

1 *where two right-angles are 360 degrees*: Ordinarily, 360 degrees equals four right angles; this convention is suitable for central angles because four right angles is the greatest central angle that can be drawn in a circle. But Ptolemy occasionally adopts an alternative convention in which 360 degrees equals *two* right angles; this is appropriate for circumferential angles, since the largest circumferential angle that can be drawn in a circle is, indeed, two right angles [Euc. III.16, 20]. Some writers use the term "demi-degree" to name the 360th part of two right angles.

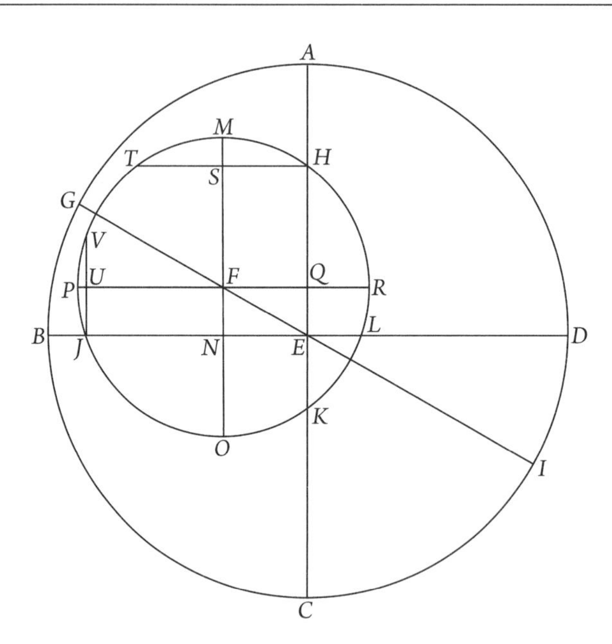

So that also, the sun will appear to traverse arc *CD*, which is from an autumnal equinox to a winter tropic, in $88\frac{1}{8}$ days, H238

and arc *DA*, which is from a winter tropic to a vernal equinox, in $90\frac{1}{8}$ days, most nearly.

And the foregoing has been found by us in agreement with what is said by Hipparchus.

Calculation 3.22

In light, then, of these quantities, let us first investigate how large the greatest difference of ⟨the arc of⟩ uniform motion compared with ⟨the arc of⟩ anomalistic motion is, and at which points such a difference will happen.

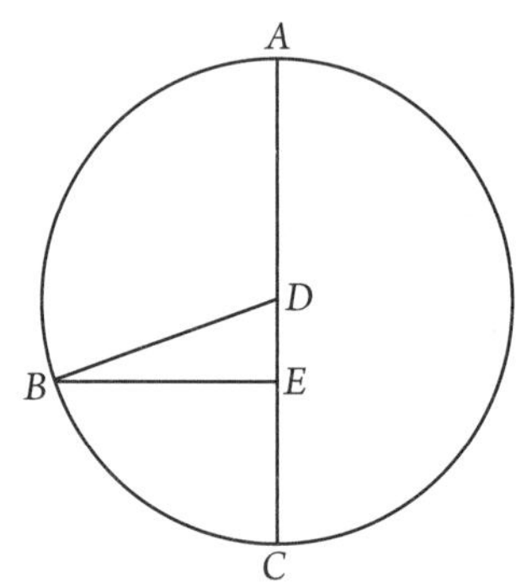

Now, let eccentric circle *ABC* be around center *D* and diameter *ADC* through the apogee *A*, upon which let *E* be the center of the zodiac circle. And let *EB* be drawn at right angles to *AC*, and let *DB* be joined.

Since the straight line *DE* between the centers is 2;30 parts where radius *BD* is 60,

by the ratio of 24 to 1, then, accordingly, straight line *DE* will be 5 parts where hypotenuse *DB* is 120,

while the arc on it will be 4;46 degrees, most nearly, where the circle around right-angled triangle *BDE* is 360 degrees.

So that also angle *DBE*, which contains the greatest anomalistic difference, will be 4;46 degrees where two right-angles are 360 degrees, and 2;23 degrees where four right-angles are 360 degrees. H239

And right angle *BED* is also 90 of the same degrees,
while angle *BDA* is equal to the two, clearly 92;23 degrees.

And since angle *BDA* is at the center of the eccentric circle and angle *BED* at the center of the zodiac circle, we will have the greatest anomalistic difference as 2;23 degrees.

And as for the arcs on which this occurs, we will have the uniform arc of the eccentric circle as 92;23 degrees from apogee, but the apparent, anomalistic arc of the zodiac circle as 90 degrees of the quadrant, just as we proved previously. It is clear from what has been treated previously that—on the opposite segment—the apparent, mean passage and the greatest anomalistic difference will be at 270 degrees, while the passage that is uniform and on the eccentric circle will be at 267;37 degrees.

Calculation 3.23

In order, as we said {H220.10}, to show numerically that the same quantities are obtained on the epicyclical hypothesis also, when the same ratios, in the manner we have stated {H219.25}, are maintained, let circle *ABC* concentric with the mid-zodiac circle be about center *D* and diameter *ADC*, and epicycle *EFG* around center *A*. And let there be

H240

drawn from *D* the straight line *DFB* tangent to the epicycle, and let *AF* be joined.

E
A
B
F
G
D
C

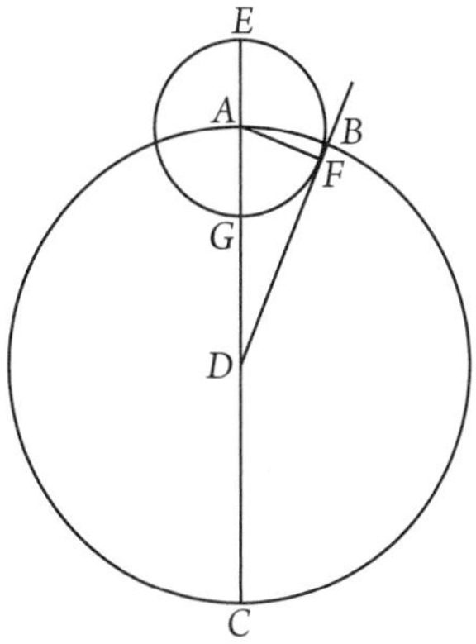

Now, in like manner {H219.21}, in right-angled triangle *ADF* side *AD* is 24 times side *AF*,
so that also, *AF* is again 5 parts where the hypotenuse *AD* is 120,
while the arc on it is 4;46 degrees where the circle described about right-angled triangle *ADF* is 360 degrees.

And therefore angle *ADF* will be 4;46 degrees where two right-angles are 360 degrees, and 2;23 degrees where four right-angles are 360 degrees.

Therefore, the greatest anomalistic difference, that is to say arc *AB*, has been found on this hypothesis as well in agreement as 2;23 degrees, and the arc of anomalistic motion, since it is contained by the right-angle *AFD*, is 90 degrees, while the arc of uniform motion, contained by angle *EAF*, again is 92;23 degrees.

5. On the Inquiry into Particular Segments of Anomalistic Motion.

In order that we may determine particular anomalistic motions on each occasion, we will again show on each of the hypotheses how, when one of the stated arcs is given, we might obtain the rest as well.

Calculation 3.24

Now, first let circle *ABC*, concentric with the zodiac circle, be around center *D*, eccentric circle *EFG* around center *H*, diameter *EAHDG* through both centers and apogee *E*, and, when arc *EF* is intercepted, let *FD* and *FH* be joined.

H241

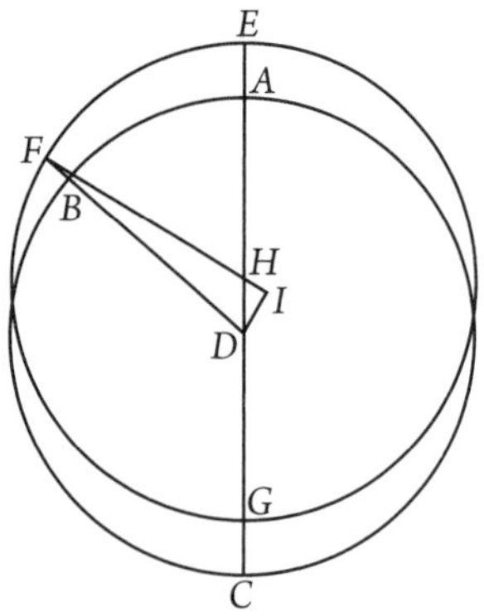

First let arc *EF* be given, for the sake of argument, as 30 degrees, and, when *FH* is produced, let *DI* be drawn from D perpendicular to it.

Since, then, arc *EF* is assumed as 30 degrees, therefore, angle *EHF*, that is to say, angle *DHI*, is 30 degrees where four right-angles are 360 degrees, and 60 degrees where two right-angles are 360 degrees.

And therefore the arc on *DI* is 60 degrees where the circle around right-angled triangle *DHI* is 360 degrees,
while the arc on *IH* is the 120 degrees supplementary to the semicircle [Euc. 3.31].

And therefore as for the chords that subtend them: *DI* will be 60 parts where hypotenuse *DH* is 120 parts, and *IH* 103;55 of the same parts.

So that also, where straight line *DH* is 2;30 parts and radius *FH* 60 parts, straight line *DI* will be 1;15 parts, *HI* 2;10 of the same parts,
and the whole *IHF* 62;10 parts.

H242

And since the squares on them when added together make the square on *FD* [Euc. 1.47],
hypotenuse *FD* will be 62;11 such parts, most nearly.

And therefore the straight line *DI* will be 2;25 parts where *FD* is 120 parts,
while the arc on it will be 2;18 degrees where the circle around right-angled triangle *FDI* is 360 degrees.
So that also, angle *DFI* is 2;18 degrees where two right-angles are 360 degrees, and 1;9 degrees where four right-angles are 360 degrees.

Therefore, the anomalistic difference at that time is 1;9 degrees.

Angle *EHF* was 30 of the same;
and therefore the remaining angle *ADB* [Euc. 1.32], that is to say, arc *AB* of the zodiac circle, is 28;51 degrees.

Calculation 3.25

The fact that, even if some other of these angles is given, the remaining angles will also be given will be immediately clear, if, in the same diagram, *HJ* is drawn perpendicular to *FD* from *H*.

For if we assume arc *AB* of the zodiac circle as given, that is to say, angle *HDJ*,
the ratio of *DH* to *HJ* will also be given through this.

And as the ratio of *DH* to *HF* is given, then the ratio of *HF* to *HJ* will be given,

and through this we will have as given angle *HFJ*, that is to say the anomalistic difference,
H243
and angle *EHF*, that is to say, arc *EF* of the eccentric circle.

And if we assume the anomalistic difference as given, that is to say, angle *HFD*, the same things will result over again.

The ratio of *HF* to *HJ* is given through this,
and the ratio of *HF* to *HD* is also given initially,
so that there is also given the ratio of *DH* to *HJ*.

And through this is also given angle *HDJ*,
that is to say, arc *AB* of the zodiac circle,
and angle *EHF* [Euc. 1.32], that is to say, arc *EF* of the eccentric circle.

Calculation 3.26

Again, let circle *ABC*, concentric with the mid-zodiac circle, be around center *D* and diameter *ADC*, and epicycle *EFGH*, in the same ratio, around center *A*, and, when arc *EF* is intercepted, let *FBD* and *FA* be joined. And let it be assumed again that arc *EF* is 30 of the same degrees; and let *IF* be drawn perpendicular from *F* to *AE*.

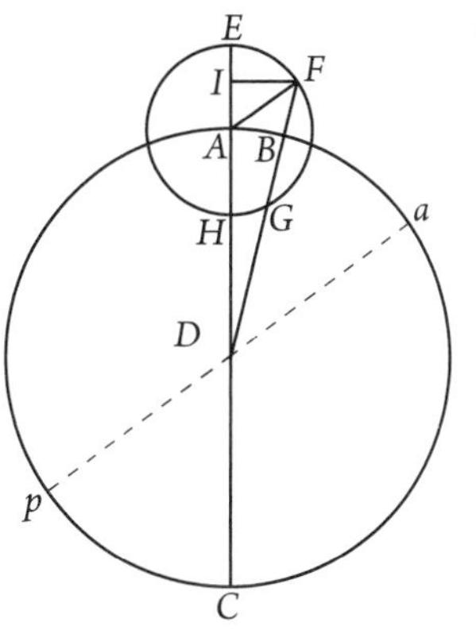

Since arc *EF* is 30 degrees,
angle *EAF* would also be 30 degrees where four right-angles are 360 degrees, and 60 degrees where two right angles are 360 degrees.

H244
So that also, the arc on *FI* is 60 degrees where the circle about right-angled triangle *AFI* is 360 degrees,
and the arc on *AI* is the 120 degrees supplementary to the semicircle [Euc. 3.31].

And therefore as for the chords subtending them:
FI will be 60 parts where diameter *AF* is 120,
and *IA* 103;55 of the same.

So that also, where hypotenuse *AF* is 2;30 parts and radius *AD* 60 parts,
straight line *FI* will be 1;15 parts,
IA 2;10 of the same parts,
and the whole *IAD* 62;10 parts.

And since the squares on them when added together make the square on *FBD* [Euc. 1.47],
FD in length will be 62;11 parts where *FI* was 1;15 parts.

And therefore straight line *FI* will be 2;25 parts where hypotenuse *DF* is 120,
while the arc on it will be 2;18 degrees where the circle around right-angled triangle *DFI* is 360.

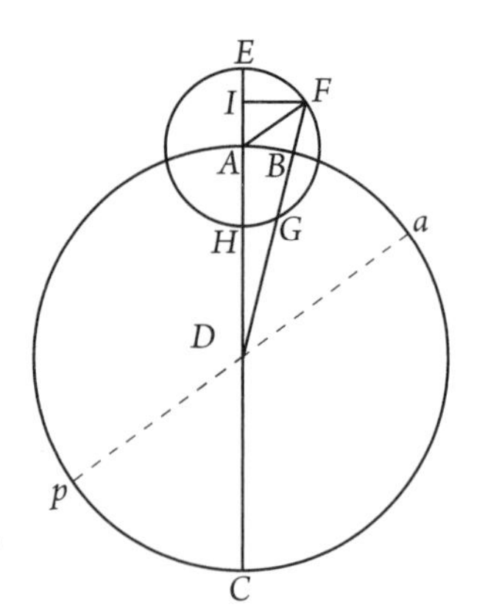

So that also, angle *FDI* is 2;18 degrees where two right-angles are 360 degrees, and 1;9 degrees where four right-angles are 360 degrees.

Therefore, the anomalistic difference, arc *AB*, is again 1;9 degrees.

Angle *EAF* was 30 of the same degrees; therefore, the remaining angle *AFD*, that is to say, the apparent arc of the zodiac circle, is 28;51 degrees:
in agreement with the quantities proved for eccentricity.

H245

Calculation 3.27

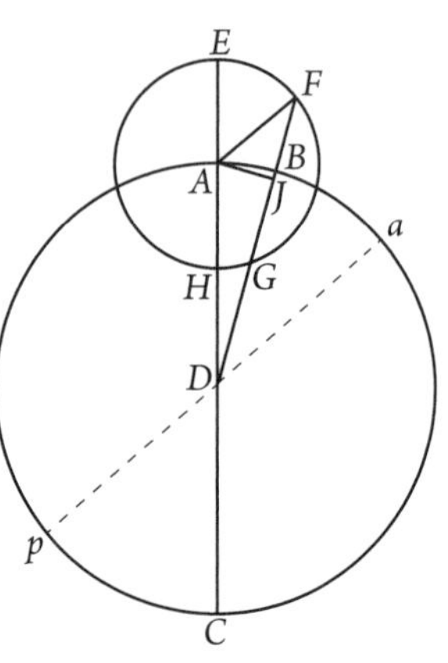

Similarly in this case too, even if another of these angles is given, the remaining angles will also be given, when, in the same diagram, *AJ* is drawn perpendicular from *A* to *DF*.

For, again, if we give the apparent arc of the zodiac circle, that is to say, angle *AFD*,
through this will be given the ratio of *FA* to *AJ*.

And as the ratio of *FA* to *AD* is also initially given,
then the ratio of *DA* to *AJ* will be given.

And, through this also, angle *ADB*, that is to say, arc *AB* of the anomalistic difference, and angle *EAF* [Euc. 1.32], that is to say, arc *EF* of the epicycle, will be given.

And if we assume the anomalistic difference as given, that is to say, angle *ADB*,
again in like manner through this will also be given the ratio of *AD* to *AJ*.

And as the ratio of *DA* to *AF* also is initially given,
the ratio of *FA* to *AJ* will also be given.

H246

Through this also, angle *AFD*, that is to say, the apparent arc of the zodiac circle, and angle *EAF*, that is to say, arc *EF* of the epicycle, will be given.

Calculation 3.28

Again, in the preceding diagram of the eccentric circle ⟨see on next page⟩, let there be intercepted from perigee *G* of the eccentric circle arc *GF* assumed as 30 of the same degrees,[1] and let *DFB* and *FH* be joined, and let *DI* be drawn from *D* perpendicular to *HF*.

Since arc *FG* is 30 degrees, angle *FHG* also would be 30 degrees where four right-angles are 360 degrees, and 60 degrees where two right-angles are 360 degrees.

1 *the same degrees*: that is, degrees measured around the center of the eccentric circle.

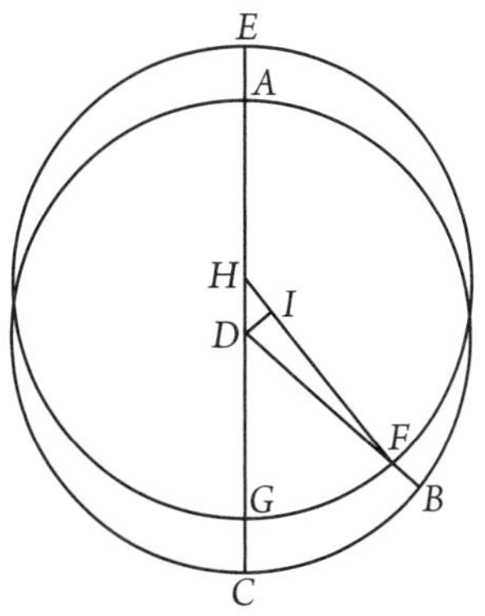

So that also the arc on straight line *DI* is 60 degrees where the circle about right-angled triangle *DHI* is 360 degrees,
and the arc on straight line *IH* is the 120 degrees supplementary to the semicircle [Euc. 3.31].

And therefore as for the chords that subtend them:
DI will be 60 parts where diameter *DH* is 120 parts,
and *IH* 103;55 of the same.

And therefore, where hypotenuse *DH* is 2;30 parts and radius *HF* 60 parts,
straight line *DI* is 1;15 parts,
H247 *HI* similarly 2;10 parts,
and *IF* the remaining 57;50 parts.

And since the squares on them when added together make the square on *DF* [Euc. 1.47],
DF will be 57;51 parts in length, most nearly, where *DI* was 1;15 parts.

And therefore *DI* will be 2;34,36 parts where hypotenuse *DF* is 120 parts,
while the arc on it will be 2;27 degrees where the circle about right-angled triangle *DFI* is 360 degrees.

So that also, angle *DFI* is 2;27 degrees where two right-angles are 360 degrees, and 1;14 degrees, most nearly, where four right-angles are 360 degrees.

Therefore, the anomalistic difference is 1;14 degrees.

And since angle *FHG* is assumed as 30 of the same degrees,[1]
the whole angle *BDC*, that is to say, arc *CB* of zodiac circle, also will be 31;14 degrees.

Calculation 3.29

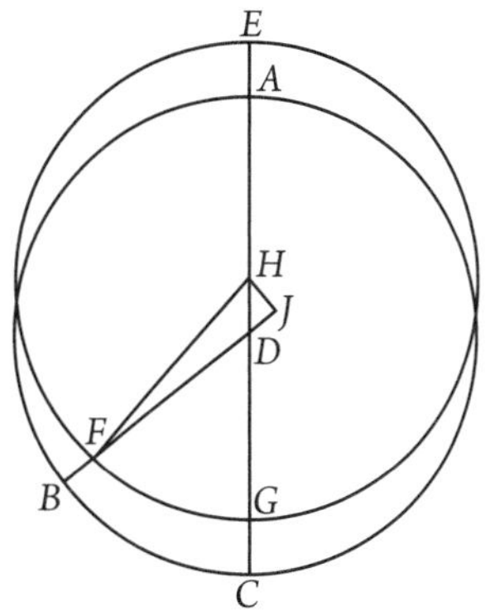

By the same considerations in this case also,
when *BD* is produced and *HJ* is drawn perpendicular to it,
if we give arc *CB* of the zodiac circle, that is to say, angle *HDJ*,
through this the ratio of *DH* to *HJ* will be given;
H248 and, as the ratio of *HD* to *HF* is also initially given,
the ratio of *FH* to *HJ* also will be given.

1 *the same degrees*: that is, full-sized degrees.

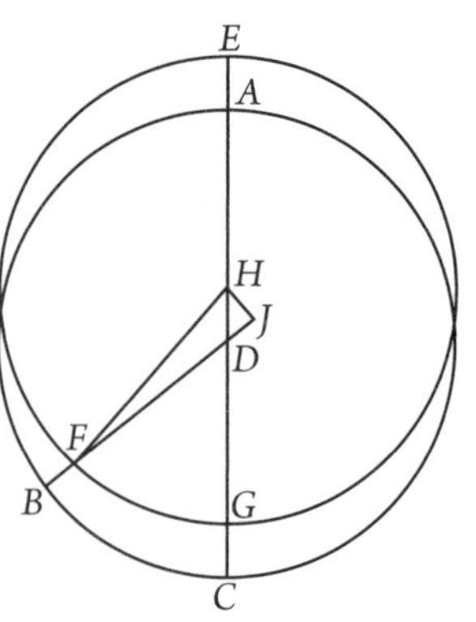

Through this we will have as given angle *HFD*, that is to say the anomalistic difference, and angle *FHD* [Euc. 1.32], that is to say, arc *GF* of the eccentric circle.

And if we give the anomalistic difference, that is to say, angle *HFD*,
once again through this the ratio of *FH* to *HJ* also will be given;
and, as the ratio of *FH* to *HD* is initially given,
the ratio of *DH* to *HJ* also will be given.

Through this we will have as given angle *HDJ*, that is to say, arc *CB* of the zodiac circle,
and angle *FHG* [Euc. 1.32], that is to say, arc *GF* of the eccentric circle.

Calculation 3.30

In like manner, in the previous diagram of the concentric circle and epicycle, when arc *HG* is intercepted from perigee *H* as 30 of the same degrees, let *AG* and *DGB* be joined, and let *GI* be drawn perpendicular from *G* to *AD*.

H249

Accordingly, since arc *HG* is again 30 degrees, angle *HAG* also would be 30 degrees where four right-angles are 360 degrees, and 60 where two right-angles are 360 degrees.

So that also, the arc on *GI* is 60 degrees where the circle about right-angled triangle *GIA* is 360 degrees,

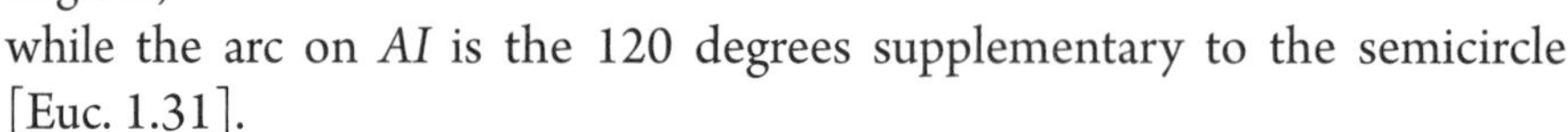

while the arc on *AI* is the 120 degrees supplementary to the semicircle [Euc. 1.31].

And therefore as for the chords that subtend them:
GI will be 60 parts where hypotenuse *AG* is 120 parts,
and *AI* will be 103;55 of the same parts.

And therefore, where straight line *AG* is 2;30 parts and radius *AD* 60 parts,
GI will be 1;15 parts,
AI, similarly, 2;10 parts,
and *ID* the remaining 57;50 parts.

And since the squares on them when added together make the square on *DG* [Euc. 1.47],
therefore *DG* itself in length will be 57;51 parts, most nearly, where chord *IG* was 1;15 parts.

And therefore chord *GI* will be 2;34,36 parts where hypotenuse *DG* is 120 parts, while the arc on it will be 2;27 degrees where the circle around right-angled triangle *DGI* is 360 degrees.

So that also, angle *GDI* is 2;27 degrees where two right-angles are 360 degrees, and 1;14 degrees, most nearly, where four right-angles are 360 degrees. H250

Therefore, the anomalistic difference, that is to say, arc *AB*, in this case also is 1;14 degrees.

And since angle *IAG* is assumed as 30 of the same,
the whole angle *BGA*, which contains the apparent arc of the zodiac circle, also will be 31;14 degrees:
in agreement with the quantities for the eccentric circle.

Calculation 3.31

By the same considerations in this case also,
when *AJ* is drawn perpendicular to *DB*,
if we give the arc of the zodiac circle, that is to say, angle *AGJ*,[1]
there will be given through this the ratio of *GA* to *AJ*.

And as the ratio of *GA* to *AD* is also initially given,
the ratio of *DA* to *AJ* will be given.

Through this we will have as given angle *ADB*,
that is to, say arc *AB* of the anomalistic difference,
and angle *HAG* [Euc. 1.32], that is to say, arc *HG* of the epicycle.

And if, again, we give arc *AB* of the anomalistic difference, that is to say, angle *ADB*,  H251
once again in like manner through this the ratio of *DA* to *AJ* will be given;

And as the ratio of *DA* to *AG* is initially given,
the ratio of *GA* to *AJ* also will be given.

Through this we will have as given angle *AGJ*, that is to say, the arc of the zodiac circle,
and angle *HAG*, that is to say, the arc of the epicycle.

And what was proposed has been proved by us.

Now, a complex table of divisions that contain determinations based on the anomalistic motion of the apparent passages can be constructed through these theorems to obtain readily the quantities of particular corrections. But we find more pleasing a table with the anomalistic differences set out alongside uniform arcs, both because of its conformity with the hypotheses

---

[1] *the arc of the zodiac circle, that is to say, angle AGJ*: Angle *AGJ* is equal to angle *pDG*, which is the arc on the zodiac circle from perigee *p* to the star at *G*.

and because of its simplicity and ready applicability for each individual calculation.

Hence, following the first theorems, set out both for numeric values and for particular segments, we geometrically calculated, just as in what we proved, the differences of anomalistic motion that correspond to each of the uniform arcs. And in general we divided the quadrants at apogees, both for the sun and for the remaining stars, into 15 segments, so that they are set out side-by-side at intervals of 6 degrees, and the quadrants at the perigees into 30 parts, so that they too are set out side by side at intervals of 3 degrees; because anomalistic differences that correspond to equal segments differ at a greater rate at perigees than at apogees.[1]

H252

Accordingly, we will arrange a table of the anomalistic motion of the sun in 45 rows again, and 3 columns, of which the first two contain the numeric values of the 360 degrees of uniform motion;[2] the first 15 rows contain the two quadrants at apogee, while the remaining 30 those at perigee. The third column contains the degrees of the addition-subtraction[3] of the anomalistic difference that correspond to each of the uniform numeric values. And the table is as follows.

6. Table of the Sun's Anomalistic Motion.

H253

degrees of uniform motion			
1 common values	2	3 addition-subtractions	
6	354	0	14
12	348	0	28
18	342	0	42
24	336	0	56
30	330	1	9
36	324	1	21
42	318	1	32
48	312	1	43
54	306	1	53

[1] *a greater rate at perigees than at apogees*: Since the sun's time from least apparent motion to mean motion is greater than the time from mean motion to greatest apparent motion, it follows that least apparent motion occurs at apogee and greatest at perigee.

[2] *the 360 degrees of uniform motion*: that is, longitudinal motion on the sun's eccentric circle.

[3] *addition-subtraction*: προσθαφαίρεσις.

degrees of uniform motion			
1 common values	2	3 addition-subtractions	
60	300	2	1
66	294	2	8
72	288	2	14
78	282	2	18
84	276	2	21
90	270	2	23
93	267	2	23
96	264	2	23
99	261	2	22
102	258	2	21
105	255	2	20
108	252	2	18
111	249	2	16
114	246	2	13
117	243	2	10
120	240	2	6
123	237	2	2
126	234	1	58
129	231	1	54
132	228	1	49
135	225	1	44
138	222	1	39
141	219	1	33
144	216	1	27
147	213	1	21
150	210	1	14
153	207	1	7
156	204	1	0
159	201	0	53
162	198	0	46
165	195	0	39
168	192	0	32
171	189	0	24

degrees of uniform motion			
1 common values	2	3 addition-subtractions	
174	186	0	16
177	183	0	8
180	180	0	0

7. On the Epoch for the Mean Passage of the Sun.

Since it remained to establish the epoch[1] of the uniform motion of the sun for enquiries into particular passages on each occasion, we also fashioned the following exposition. In general, we followed, again, for the sun and the other stars, the passages observed most carefully by ourselves. And from them we reckoned the establishments of epochs back to the beginning of the reign of Nabonassar based the mean motions that we are demonstrating; we possess ancient observations beginning from this time that are preserved for the most part up to the present. H254.3

Calculation 3.32

Now, let circle *ABC* concentric with the mid-zodiac circle be around center *D*, the eccentric circle of the sun *EFG* around center *H*, and diameter *EAGC* through both centers and apogee *E*. And let it be assumed that *B* is the autumnal point[2] of the zodiac circle, and let *BFD* and *FH* be joined, and let *HI* be drawn perpendicular from *H* to *FD* produced.

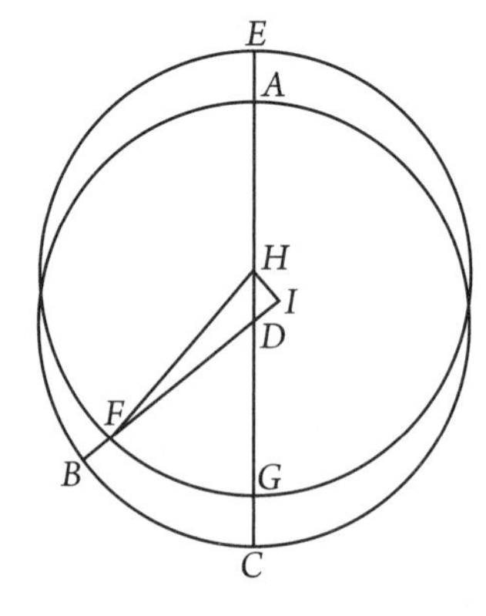

Since the autumnal point *B* contains the beginning of Libra, H255

[1] *epoch*: The Greek verb ἐπέχω, from which the noun *epoch* was formed, means, among other things, *to occupy*. In Book II {H172}, Ptolemy speaks of the "epochs" of cities, meaning their position in longitude and latitude. Thus the epoch of a star is the place on the mid-zodiac circle occupied by that star at a specified time. Noon of the first day of the reign of Nabonassar (February 26, 747 BCE) is chosen as the arbitrary starting point in time for calculating the position of all of the planets. The place (epoch) on the mid-zodiac circle (ecliptic) each of the planets occupied at that time is calculated from its known position at a later date and its mean motion, going backward, that is, westward through the zodiac. From the "epoch" one can calculate forward or backward to determine the position of the planet at any other time. The term is still used in modern astronomy.

[2] *Autumnal point*: the position of the sun at the autumn equinox, the beginning of the sign of Libra. To follow Ptolemy's argument here it will be helpful to consult a chart of the zodiac.

while perigee C contains $5\frac{1}{2}$ degrees of Sagittarius,[1]
therefore arc BC is 65;30 degrees.

And therefore angle BDC, that is to say, angle HDI, is 65;30 degrees where four right-angles are 360 degrees, and 131 where two right-angles are 360 degrees.

So that also, the arc on chord HI is 131 degrees where the circle around right-angled triangle DHI is 360 degrees,
while the chord subtending it, HI, is 109;12 parts where diameter DH is 120 parts.

Therefore HI will be 4;33 parts where straight line DH is 5 parts and hypotenuse FH 120 parts,
while the arc on it will be 4;20 degrees where the circle around right-angle triangle HFI is 360 degrees.

So that also, angle HFI is 4;20 degrees where two right-angles are 360 degrees, and 2;10 degrees where four right-angles are 360 degrees.

Angle BDC was 65;30 of the same degrees;
and therefore the remaining angle FHG, that is to say, arc FG of the eccentric circle, is 63;20 degrees.

Therefore, when the sun is at the autumnal equinox, in its mean motion it is westwards of perigee (that is to say, $5\frac{1}{2}$ degrees of Sagittarius) by 63;20 degrees,
while in its mean motion it is separated from apogee (that is to say, 5;30 degrees of Gemini) eastwards by 116;40 degrees.

H256

Observation 3.31

Now that this has been investigated, since among the first equinoxes observed by us one of those gotten most accurately, an autumnal equinox, occurred in the 17th year of Hadrian, by Egyptian reckoning Athur 7, two equinoctial hours, most nearly, after noon, it is clear that, during that time the sun in its mean motion was separated from apogee on the eccentric circle eastwards by 116;40 degrees. But from the reign of Nabonassar up to the death of Alexander 424 years (by Egyptian reckoning), are totaled up, and from the death of Alexander up to the reign of Augustus 294 years. And from the first year of the reign of Augustus (by Egyptian reckoning noon Thoth 1, since we establish epochs from noon), until the 17th year of Hadrian, Athur 7, two equinoctial hours after noon, there are 161 years, 66 days, and 2 equinoctial hours. And therefore from the first year of Nabonassar, by Egyptian reckoning mid-day Thoth 1, until the time of the above autumnal equinox,

[1] *perigee C contains $5\frac{1}{2}$ degrees of Sagittarius*: In Book III, Chapter 4, it was shown that the apogee lies westward of the summer tropic (solstice) by 24;30 degrees. The summer solstice is at the beginning of the sign of Cancer; thus the apogee lies at $5\frac{1}{2}$ degrees in Gemini. The perigee lies opposite this, at $5\frac{1}{2}$ degrees in Sagittarius.

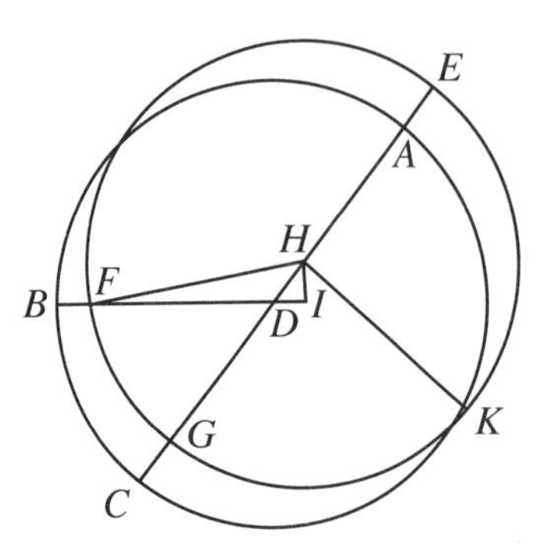

there will be totaled up 879 Egyptian years, 66 days, and 2 equinoctial hours. But in this span of time the sun in its mean motion moves, beyond whole cycles, 211;25 degrees, most nearly. Accordingly, if to the 116;40 degrees of the separation from apogee of the eccentric circle at this autumnal equinox ⟨arc *EF*⟩[1], we add 360 degrees of one cycle and from their sum we subtract 211;25 degrees of surplus for the intervening time ⟨arc *FEK*⟩, we will have for the epoch ⟨*K*⟩ of mean motion: in the first year of Nabonassar, by Egyptian reckoning Thoth 1, mid-day, the sun was separated from apogee eastwards in its uniform motion by 265;15 degrees ⟨arc *EFGK*⟩, and occupied in its mean motion 0;45 degrees of Pisces.

H257

8. On the Calculation of the Sun.

Accordingly, whenever we wish to know the passage of the sun for each time we require, totalling up the time from the epoch until the time in question, relative to the hour in Alexandria, and applying this to the tables of uniform motion, we will add successively the degrees that are set out alongside their respective ⟨time⟩ values along with 265;15 degrees of separation. And, after removing whole cycles from their sum, we will subtract the remaining degrees from the 5;30 degrees in Gemini eastwards along the zodiac signs, and, wherever the number falls out, there we will find the mean passage of the sun. Next, by applying the same number, that is to say, that from apogee up to mean passage, to the table of anomaly, we will subtract the degrees set alongside the value in the third column from the epoch for mean passage, if the number falls in the first column, that is to say, is up to 180 degrees. But if the number occurs in the second column, that is to say, falls beyond 180 degrees, we will add them to the mean passage, and in this way we will find the true, apparent ⟨position of the⟩ sun.

H258

9. On the Inequality of Days and Nights.

The subjects, then, that are investigated concerning the sun alone are roughly the preceding. It would follow as well briefly to add to them also

[1] The adjacent diagram, to which the bracketed letters refer, has been supplied for this edition. Point *B* is the autumnal equinox of 17 Hadrian, Athur 7, and therefore line *DB* is in the direction of 0 degrees of Libra.

what ought to be premised concerning the inequality of days and nights; because all the particular mean movements that have been set out simply by us receive augmentation by equal differences, as though all the days and nights are also of equal times; but this is observed not to be the case. The revolution of all the stars is produced uniformly around the poles of the equatorial circle and this sort of periodic return is taken relative either to the horizon or to the meridian circle, according to which is the more conspicuous. Therefore it is clear that the periodic return of the same point of the equatorial circle, from some segment, whether of the horizon or the meridian circle, to the same point is a single revolution of the universe. And the periodic return of the sun from some segment, whether of the horizon or the meridian circle, back to the same point is simply a day and night. For these reasons, then, a uniform day and night is that which contains a passage of 360 time-degrees of one revolution of the equatorial circle, and, furthermore, 0;59 of one time-degree, most nearly: as much as the sun in its mean motion moves onward in this amount of time. And an anomalistic day and night is that which contains a passage of the 360 time-degrees of a single revolution of the equatorial circle and, furthermore, of the time-degrees that either co-rise or co-culminate with the anomalistic, onward motion of the sun.

I259

Now, this segment of the equatorial circle that additionally passes through must be unequal to 360 time-degrees, because of the apparent anomalistic motion of the sun and because equal segments of the mid-zodiac circle do not pass through either the horizon or the meridian circle in equal times. However, each of these makes the difference of the uniform periodic return for one day and night compared with the anomalistic periodic return imperceptible, but quite perceptible when added up from many days and nights.

The greatest solar, anomalistic difference, then, occurs over intervals from one of the mean motions of the sun to the other. For days and nights when summed up in this manner will differ from the uniform days and nights by $4\frac{3}{4}$ time-degrees, most nearly, and from one another by twice the number of time-degrees, $9\frac{1}{2}$; because the apparent passage of the sun compared with its uniform passage on the semicircle at apogee falls short by $4\frac{3}{4}$ degrees, but on the semicircle at perigee is greater by the same number of degrees. And the greatest anomalistic difference due to co-risings or co-settings occurs over semicircles determined by the tropic points. For in this case also, the co-risings of each of these semicircles will differ from the 180 time-degrees, which are observed as uniform, by the differences of the longest or shortest day compared with the equinoctial day, and from one another by those differences by which the longest day or night differs from the shortest. The greatest difference due to the inequality of co-culminations occurs, again, over intervals that contain, roughly, two signs of the zodiac that are at the

H260

same time each on one side, whether of the tropic or equinoctial points. For, in fact, two of these taken together at the tropic points will differ from those observed as uniform by $4\frac{1}{2}$ time-degrees, most nearly, while from two taken together at the equinoctial points by 9 time-degrees; because the latter fall short compared with the mean value, while the former are greater by roughly the same amount. Hence we establish the epochal starting points of days and nights from culminations and not from risings or settings of the sun; because the difference that is observed relative to horizons can extend up to many hours and is not invariably the same, and it changes along with the difference among longest or shortest days at each inclination of the sphere. But the difference relative to the meridian circle is the same for every zone of habitation and does not exceed at all the time-degrees of the difference that are summed up from the sun's anomalistic motion. And from the mixture of both of these, the difference due to the anomalistic motion of the sun and the difference due to the co-culminations, is also established the difference found in the intervals that are either at the same time additive or subtractive according to the two stated differences. The segment from the middle of Aquarius up to Libra is most subtractive from both sources, but the segment from Scorpio up to the middle of Aquarius additive; because each of the stated segments either adds or subtracts the greatest amount: $3\frac{2}{3}$ degrees, most nearly, due to the sun's anomalistic motion, and $4\frac{2}{3}$ time-degrees, most nearly, due to the co-culminations. So that from the stated mixture, a greatest difference of days and nights, on each of the stated segments, sums up as $8\frac{1}{3}$ time-degrees relative to uniform days and nights, that is to say, $\frac{5}{9}$ of one hour, but relative to one another twice the number of time-degrees, $16\frac{2}{3}$, that is to say, $1\frac{1}{9}$ hours. This amount, if disregarded for the sun and the other stars, would perhaps not impair the inquiry into their appearances by anything perceptible; but for the moon, because of the speed of its motion, it would render the difference significant indeed, even up to $3\frac{1}{5}$ degrees.

H261 H262

In order, then, that we resolve once and for all days and nights given over any interval whatsoever, I mean those from mid-day or mid-night up to mid-day or mid-night, into uniform days and nights, by the prior and the subsequent epoch of the given interval of days and nights we will inquire into what degrees of the mid-zodiac circle the sun occupies while it moves both uniformly and anomalistically. Next, by applying the interval from the anomalistic, that is to say, the apparent, interval to the apparent interval of the degrees of surplus, we will inquire into ascensions in right sphere: in how many time-degrees of the equatorial circle do the degrees of the anomalistic interval, as we said {H259.7–11}, co-culminate. And we will take the difference between the time-degrees we find and the degrees of the uniform interval and we will calculate the magnitude of the equatorial hour that is

H263

comprised by the time-degrees of the difference. When the number of time-degrees of the uniform interval is found greater, we will add this magnitude to the given quantity of days and nights; but when it is less, we will subtract this magnitude from it. And we will have the resultant time divided into uniform days and nights; this we will employ, above all, for the additions of mean motions in the tables of the moon. And it is immediately easy to understand that, from the foundation of uniform days and nights, days and nights that vary by the season and are observed plainly are also obtained, if the preceding addition-subtraction of the hourly time-degrees is reversed.[1]

According, then, to our epoch, that is to say in the first year of Nabonassar, by Egyptian reckoning Thoth 1, mid-day, in its uniform motion, as we proved shortly above {H257.6}, the sun occupied 0;45 degrees of Pisces, but in its anomalistic motion occupied 3;8 degrees, most nearly, of Pisces.

---

[1] The difference between uniform days and the nonuniform solar days has a twofold origin: the obliquity of the mid-zodiac circle and the nonuniform motion of the sun.

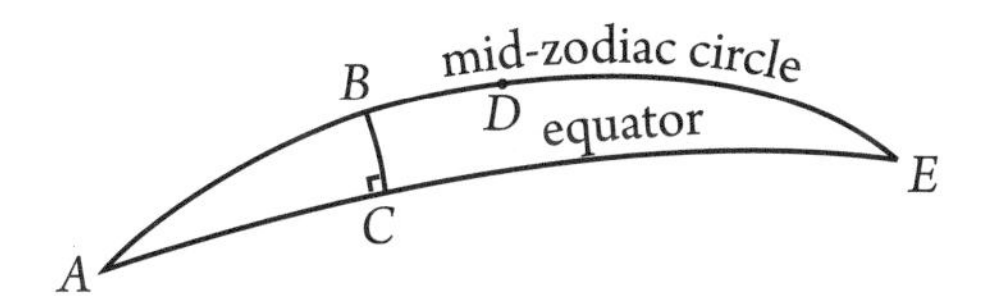

To understand the first, imagine that the sun is moving uniformly along the mid-zodiac circle. According to Ptolemy's calculations 1.34–1.44 {H85}, after the first ten degrees of its motion from the spring equinox (arc *AB*), the sun's motion as measured along the equator (arc *AC*) will have fallen behind the uniform rate on the equator by 50 arc minutes, and the sun will therefore require slightly more time to arrive at the meridian. Therefore, at this time of year the solar days would be slightly longer than the uniform days. This deficit would increase up to about the middle of the spring, after which it would catch up to the uniform motion on the equator at the summer solstice (*D*). From summer solstice to autumn equinox (arc *DE*), it would first move faster than the uniform motion, making the days slightly shorter, but then would fall back, becoming again equal to the uniform motion at the equinox (*E*). The pattern would repeat over the second half of the year.

But the sun does not move uniformly along the mid-zodiac circle; rather, it goes a little faster near perigee (in December) and a little slower near apogee (in June). Therefore, one must add or subtract this to or from the mean motion on the mid-zodiac circle, and then apply this corrected motion to the degrees on the equator (the "time degrees"). This will give the time by which the solar noon falls ahead of or behind the uniformly determined noon. The computation can be reversed to find the uniform "mean noon" from the solar noon.

Book VII

1. That the Fixed Stars Always Preserve the Same Position Relative to One Another.

H2

Having explained in the books before this, Syrus, what happens in right sphere and inclined sphere and, furthermore, what relates to the hypotheses of the motions of the sun and moon and the configurations observed among them, we will now, for the sake of the subsequent inquiry, begin our account of the stars and first in order, our account of those called "fixed."

Now, the following question must be taken up first of all. Because these stars, true to their name, clearly always preserve the same configurations and equal distances relative to one another, we could rightly call them "fixed". But because the whole sphere (on which they are carried as if grafted to it) itself also clearly makes a distinct and orderly motion to the east, or rearwards, of the primary motion, it would no longer be fitting to call this sphere "fixed" also. For we find that both these assertions are valid, at least on the basis of what the time within our purview suggests. Hipparchus still earlier surmised them both on the basis of the appearances that he had. And so, then, he made guesses rather than affirmations about a greater span of time, because he had stumbled upon very few observations of the fixed stars before himself (roughly speaking, only those recorded by Aristyllos and Timocharis, although these were neither unambiguous nor worked out). We too reach the same understanding on the basis of the comparison between what is observed now and what was observed then; though by now it is more certain, because the examination has taken place for a greater time and the records of Hipparchus on the fixed stars, with which we have above all made comparisons, have been handed down to us with total completeness.

H3

No change, then, of their relative positions has occurred up to the present; but the configurations observed by Hipparchus are even now seen indistinguishably the same. And it is not only those in the zodiac circle relative to one another or those outside it relative to those likewise outside of it; (this would result, if, on the first hypothesis that Hipparchus sets forth, only the stars around the zodiac circle itself were to move eastwards). But it is also the case with the configurations of the stars in the zodiac circle relative to those outside it and farther removed. These facts would prove easy to understand by anyone who wishes to advance the inquiry and examine carefully, with a love for truth, whether present appearances are in agreement with Hipparchus's records.

H4

Accordingly, we will set out here as well, by way of a ready test, a few records that can be most easily understood and can bring the entire comparison into view, because they show that the observed configurations that are comprised by the stars outside the zodiac are the same relative to one another and to those within the zodiac circle.

Observation 7.1

For the stars in Cancer, then, Hipparchus records that the star on the southern claw of Cancer, the bright star to the west of this claw and the head of Hydra, and the bright star in Procyon are, most nearly, in a straight line.[1] For the middle one of them diverges from the straight line through the outermost stars, to the north and east, by $1\frac{1}{2}$ digits, while the distances between them are equal.

[The rest of Hipparchus's observations are omitted, H4.17–5.7]

H8.8

Now of these and such configurations that admit of comparison through nearly the whole of the sphere, we see none that has altered up to the present time; this would have resulted in a quite perceptible manner in the nearly two hundred and sixty intervening years, if those stars around the zodiac circle alone were to move eastwards.[2]

[Ptolemy's own observations are omitted, H.8.15–12.3]

2. That the Sphere of the Fixed Stars Also Makes a Certain Motion Eastwards Through the Mid-zodiac Circle.

H12.4

We can establish, then, that the disposition and motion of all the so-called fixed stars without exception is one and the same from these and similar observations. But it has become clear to us that their sphere also makes a certain distinct motion in the direction opposite to the movement of all the stars (that is to say eastwards of the great circle described through the poles of both circles, the equatorial and the mid-zodiac circle), principally because the same stars do not preserve the same distances, both long ago

[1] *The stars in Cancer ... the bright star in Procyon are, most nearly, in a straight line*: The drawing shows the stars in question. Star α in Cancer is "the star in the southern claw"; the star "west of this claw and the head of Hydra" is β.

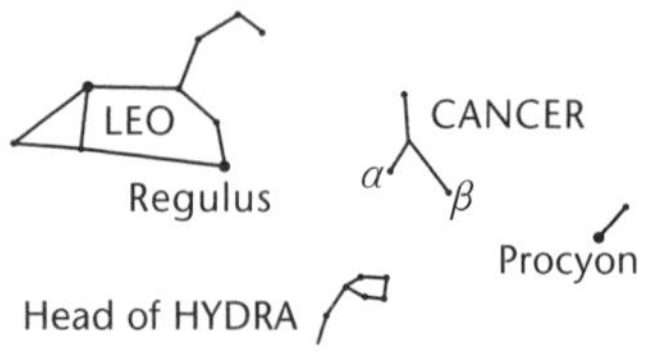

[2] *if those stars around the zodiac circle alone were to move eastwards*: Since Cancer, for example, lies within the zodiac while Procyon does not, the relative positions of the three stars noted above would have changed if there were a motion confined to the zodiac.

and in our time, relative to the tropic and equinoctial points,[1] but are always found, the later the times, to be removed a greater distance eastwards of these same points than their preceding distance.

For Hipparchus, in his *On the Change of the Tropic and Equinoctial Points*, in laying out lunar eclipses on the basis of those he himself accurately observed and those yet earlier Timocharis observed, calculates that in his own times Spica was removed 6 degrees westwards from the autumnal point while in the times of Timocharis 8 degrees, most nearly. For in concluding he says,

H13

> Therefore, if, for example, Spica was to the west in longitude of the autumnal point of the zodiac signs by 8 degrees earlier, but now is to the west by 6 degrees,

and all his additional remarks. And for the rest of the fixed stars that he has compared, he proves that an eastward recession[2] of roughly this amount has occurred. And, comparing the apparent distances in our own times of the fixed stars relative to the tropic and equinoctial points to those both observed and recorded by Hipparchus, we find no less that their eastward through the mid-zodiac circle has occurred in proportion with his proposed motion. (We have made this an examination through the instrument that was previously constructed by us for careful observations of the particular distances of the moon from the sun. We station one of the rings of the astrolabes towards the apparent passage of the moon taken at the hour of observation, while we move the other towards the star being sighted, in order that the moon and the star be observed simultaneously at their proper places. And in this way we obtain, from the distance relative to the moon, the position of each one of the bright stars as well).

H14

Observation 7.2

For, to take one example, we observed in the second year of Antoninus, by Egyptian reckoning Pharmouthi 9th, the sun being about to set at Alexandria, while the final segment of Taurus was culminating[3] (that is to

[1] Ptolemy treats the zodiac signs as fixed relative to the tropic and equinoctial points, while he finds that the zodiac constellations move relative to these points. Thus, for example, the vernal equinox always occurs (by convention) at the beginning of the sign of the Ram. In Ptolemy's time, the constellation Aries was located in this sign, but now the constellation Aries appears eastward, in the sign of the Bull.

[2] *recession*: Since Ptolemy regards the solstice and equinoctial point as fixed, he sees the relative motion of the constellations and these points as a "recession" (eastward motion) of the constellations. If the constellations are regarded as fixed, the same relative motion would be described as a "precession" (westward movement) of the equinoxes. See the remark on precession of the equinoxes in the *Preliminaries to Book III*, pages 77–78.

[3] *culminating*: A celestial body is said to culminate when it reaches its highest point in the sky, that is, when it crosses the local meridian.

say, $5\frac{1}{2}$ equinoctial hours after noon on the 9th) that the apparent moon was removed $92\frac{1}{8}$ degrees from the sun, sighted about 3 degrees of Pisces. Half an hour later, when the sun had already set and the first quarter of Gemini was culminating and the apparent moon was sighted in the same position, through one of the astrolabes the star on the heart of Leo[1] appeared removed from the moon eastwards, again, by $57\frac{1}{6}$ degrees on the mid-zodiac circle. But previously, the sun occupied precisely $3\frac{1}{20}$ degrees, most nearly, of Pisces, so that at that time the apparent moon occupied $5\frac{1}{6}$ degrees, most nearly, of Gemini, because of its distance of $92\frac{1}{8}$ degrees eastwards: as many degrees as it ought to have occupied on our hypotheses also. One half hour later, the moon ought to have advanced $\frac{1}{4}$ degree, most nearly, eastwards and have been displaced westwards $\frac{1}{12}$ degree,[2] most nearly, in comparison with its first position. One half an hour later, then, the apparent moon occupied H15 $5\frac{1}{3}$ degrees of Gemini, so that the star on the heart of Leo, since it was clearly removed $57\frac{1}{6}$ degrees from it eastwards, also occupied $2\frac{1}{2}$ degrees of Leo, and was removed $32\frac{1}{2}$ degrees from the summer tropic point.[3]

But in the 50th year of the third Calippic cycle, as Hipparchus who observed it records, the star on the heart of Leo was removed $29\frac{5}{6}$ degrees from the same summer tropic point again eastward. Therefore, the star on the heart of Leo was displaced $2\frac{2}{3}$ degrees eastwards on the mid-zodiac circle. The years from Hipparchus' observation up to the beginning of the reign of Antoninus, during which especially we have observed the greatest number of passages of the fixed stars, sum up to about 265. So that from these considerations the displacement of one degree eastward has been found to have occurred in one hundred years, most nearly, just as Hipparchus had clearly suspected, according to what he says in his *On the Annual Period*:

> For if both the tropics and equinoxes were, due to this cause, moving westwards[4] through the zodiac signs in the span of a year not less than $\frac{1}{100}$ degree, they ought to have moved no less than 3 degrees in three hundred years. H16

---

[1] "The star on the heart of Leo" is Regulus, shown in the drawing. Besides being a very bright star, Regulus is especially useful for locating celestial positions because it lies nearly on the mid-zodiac circle.

[2] *the moon ought to have advanced* 1/4 *degree ...* 1/12 *degree*: Ptolemy derives these corrections on the basis of his theory of the moon's motion outlined in Book V, omitted in these selections.

[3] The summer tropic, by convention, lies at the beginning of the sign of the Crab.

[4] Hipparchus apparently sees the relative motion as a precession of the equinoxes.

And once we sight, in the same way, Spica and the brightest stars of those around the mid-zodiac circle from the moon, and then subsequently more easily sight the rest of the stars from these stars, we find that their mutual distances again are the same, most nearly, as those observed by Hipparchus. But we find that their distances relative to the tropic and equinoctial points in each case have been displaced $2\frac{2}{3}$ degrees, most nearly, eastwards compared with the record of Hipparchus.

3. That the Eastward Motion of the Sphere of the Fixed Stars is Also Produced Around the Poles of the Mid-zodiac Circle.

That the sphere of the fixed stars, then, also makes a motion of this amount, most nearly, eastwards through the mid-zodiac circle has become easily understood by us through these observations. It follows to inquire into the mode of such a motion, that is to say, whether it is produced around the poles of the equatorial circle or those of the oblique, mid-zodiac circle. This would become clear from the longitudinal recession itself, since great circles described through the poles of one of the stated circles cut off unequal arcs on the other,[1] unless the difference due to the previously mentioned cause happened still to be utterly imperceptible, since the longitudinal in this span of time was small. But this would be easiest to understand through their latitudinal passage of old and now. For it is clear that the motion of the sphere of the fixed stars will be produced around the poles of whichever of the two circles, the equatorial or the mid-zodiac, in relation to which they always clearly preserve their latitudinal distance.

H17

Hipparchus, then, gives his assent to the motion that occurs around the poles of the oblique circle. For, in his *On the Change of the Tropic and Equinoctial Points* again, he concludes, on the basis of what Timocharis observed and what he himself observed, that Spica has preserved the magnitude of its latitudinal distance not in relation to the equatorial circle but rather to the mid-zodiac circle, and is 2 degrees more southerly than the mid-zodiac circle, both earlier and later. And, for this reason, in his *On the Annual*

[1] *great circles desribed through the poles of one of the stated circles cut off unequal arcs on the other*: In the drawing, point *A* is the vernal equinox. The meridian lines *NAS, NDBS, NECS* (drawn through the poles of the equator) define two equal arcs *AB* and *BC* along the equator, while also cutting off arcs *AD* and *DE* on the mid-zodiac circle. The resulting spherical triangles *ADB* and *AEC* have right angles at *B* and *C*. It follows from the spherical Menelaus Theorems (Appendix 3) that if *AB* equals *BC*, *AD* cannot equal *DE*.

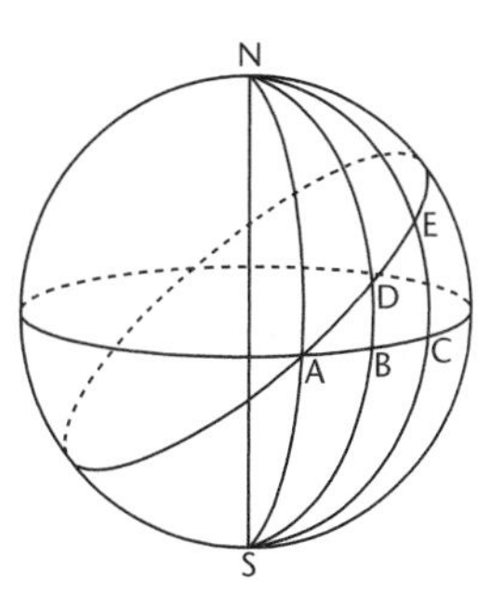

Period, he posits only the motion that occurs around the poles of the mid-zodiac circle. Nevertheless, he still is in doubt, just as he himself says, because neither are the observations, which were taken quite roughly, of the followers of Timocharis credible, nor is the difference over the intervening time quite sufficient to grasp with certainty. We, however, finding that this has been observed over a still greater time-span and for nearly all of the fixed stars, would now rightly believe that motion around the poles of the oblique circle is more certain. For in observing the latitudinal distances of each, relative to the mid-zodiac circle, on the great circle described through its poles we find them roughly the same as those recorded and calculated by Hipparchus or differing by the slightest amount and as much as could be overlooked due to the observations themselves. As for the distances observed, relative to the equatorial circle, on a great circle described through its poles, we find that the distances taken by us are not in agreement with those recorded in the same way by Hipparchus, nor are these in agreement with those recorded yet earlier by the followers of Timocharis. But we also find the sameness of their latitude relative to the mid-zodiac circle is established still further from the following consideration. The stars in the hemisphere from the winter tropic over the vernal point up to the summer tropic are always found more northerly than their more ancient distance relative to the equatorial circle, and the stars in the opposite hemisphere more southerly. And those near the equinoctial points are always found more northerly or southerly by greater differences, while those near the tropic points by smaller differences; that is, roughly speaking, the differences by which the eastern segments of the mid-zodiac circle are more northerly or southerly than the equatorial circle in their proportional, longitudinal recession.[1]

H18

H19

[The rest of Chapter 3 is omitted, H19.8–34.8]

[1] A table comparing ancient and modern zodiac signs and constellations appears in the *Preliminaries to Book III*, page 78.

PRELIMINARIES TO BOOK IX

Mercury and Venus

1. *Appearance and motion in the heavens.* Mercury is ordinarily visible for only two two-week periods each year, near the horizon at sunrise or sunset. Venus too alternates between sunrise and sunset appearances, but it remains visible for longer periods than does Mercury and is far brighter, being surpassed in brilliance only by the sun and the moon.

Like the sun, Mercury and Venus move generally eastward through the zodiac. Each is always found within a characteristic distance from the sun—about 28 degrees for Mercury, 47 degrees for Venus. Each of them, therefore, completes one cycle about the zodiac in the same average time that the sun does.[1] Their apparent motions, however, are even more irregular than is the sun's. Venus, for example, may sometimes pass twice as quickly through one zodiac sign as it does through another.

Each of these planets appears sometimes as a morning star, rising in the east shortly before sunrise, and sometimes as an evening star, setting in the west shortly after sunset. The lefthand diagram shows Venus, *V*, as morning star; it rises at *R* before the sun rises at *S*. The righthand diagram depicts Venus as evening star; it sets at *R* after the sun sets at *S*.

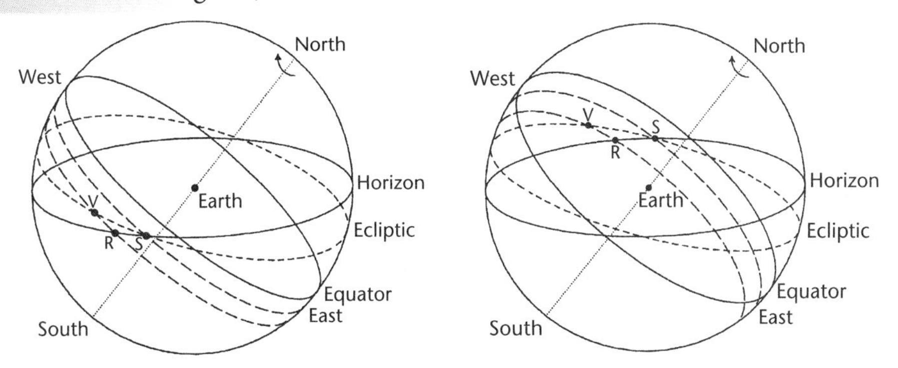

Because of its proximity to the sun, Mercury is hardly ever visible for more than 60 or 70 minutes before sunrise or after sunset. Venus, by contrast, may sometimes be viewed as much as $3\frac{1}{2}$ hours before sunrise or after sunset.

---

[1] The basis for this inference is discussed in the section "The Planets and the Mean Sun," below in these Preliminaries.

It is not obvious by sight alone that a planet's morning and evening appearances are in fact manifestations of the same planet. The Greeks in Homer's time regarded Venus the morning star and Venus the evening star as two different planets, which they named *Phosphoros* and *Hesperos*, respectively. By the time of Plato, however, it was recognized that these two were actually one planet, periodically rendered invisible by the light of the sun.

2. *Elongation and retrogradation.* When either Mercury or Venus emerges as a morning star from a period of invisibility (its *apparition*) it is west of, and very close to, the sun—so that it rises in the morning just before the sun does; it is said to exhibit a *western elongation,* measured in degrees along the ecliptic. In subsequent days its western elongation grows larger and larger until it reaches a maximum—the *greatest western elongation,* abbreviated GWE. The elongation then decreases until the planet becomes hidden by the sun's light. After a period of invisibility it reappears close to the sun, but now as an evening star, and therefore east of the sun along the ecliptic; it is said to exhibit an *eastern elongation.* In subsequent days its eastern elongation increases to a maximum—a *greatest eastern elongation* or GEE—after which the elongation again decreases until the planet is once more rendered invisible by the sun and the whole cycle begins anew.

At some time after reaching a GEE, but before attaining the following GWE, the planet appears to stop its eastward motion with respect to the ecliptic; it is then said to have reached a *station.* It next begins to move westward along the ecliptic; this western movement is called *retrogradation.* Retrogradation continues as the planet's eastern elongation decreases to the point of occultation, and continues further as the planet re-emerges with an increasing western elongation. Prior to attaining greatest western elongation, however, retrogradation ceases: the planet exhibits a second station and then resumes its eastward motion along the ecliptic.

It is an additional complication that the greatest eastward and greatest westward elongations are themselves variable; as we shall see, they depend on the planet's position along the ecliptic. Because of the sun's brightness, it is hardly possible actually to observe planetary elongations with respect to the apparent sun. Ptolemy instead expresses elongations with respect to the calculated *mean sun.*

Mars, Jupiter, and Saturn

3. *Appearance and motion in the heavens.* Mars is famously reddish in color. It varies considerably in brilliance, sometimes approaching the brightness of Jupiter, which in turn is brighter than any star other than Venus. Saturn shines about as brightly as do the brightest among the fixed stars.

Like the other planets, these three planets move generally eastward along the zodiac. On the average, Mars completes one cycle about the zodiac in about two years, Jupiter in about twelve years, and Saturn in about twenty-nine years. Nevertheless their apparent rates of progress can vary greatly. For example, Mars may move through a single zodiac sign in as little as one month or as long as several months.

4. *Elongation and retrogradation.* Unlike Mercury and Venus, the planets Mars, Jupiter, and Saturn can appear at any angular distance from the sun; they never exhibit a greatest elongation. Thus at any time of the year each of them can be anywhere on the ecliptic. When these planets are near the sun they are said to be in *conjunction,* and are then difficult or impossible to see. These periods of invisibility may last for several months.

When one of these planets is opposite the sun on the ecliptic it is said to be in *opposition.* It then rises when the sun is setting and sets when the sun is rising. When the planet is near opposition, it appears to cease its eastward motion through the zodiac; as with Mercury and Venus, this cessation of motion is called a *station.* Following this station, the planet reverses its motion and travels westward along the ecliptic; as before, this westward motion is termed *retrogradation.* The planet then reaches a second station, after which it resumes its eastward motion through the zodiac.

In the case of Mars, the appearance of station may last for as long as two weeks; retrogradation may last from about 40 to about 90 days, depending on where it is in the zodiac. All three planets, but especially Mars, appear noticeably brighter when they are in retrograde motion.

Planetary Anomalies

5. *Zodiacal and heliacal anomaly.* A celestial body's departure from uniform motion is termed an anomaly. In the case of the sun we observe only a single anomaly: the sun's position on the ecliptic sometimes exceeds, sometime falls short of where it would be if it always moved uniformly. Furthermore, the magnitude of this discrepancy depends on the time of year; that is to say, it depends on where the sun is in the zodiac. Because of its connection with zodiacal position, the sun's anomaly is considered a *zodiacal* anomaly.

In the case of the planets, Ptolemy identifies two anomalies. One, like the sun's, depends on the part of the zodiac in which the planet appears and is therefore a zodiacal anomaly. The other relates to the planet's position, not with respect to the zodiac but with respect to the sun; it is therefore called a *heliacal* anomaly. As Ptolemy acknowledges in Chapter 2 below, the two anomalies are intermixed and thus difficult to distinguish. The following tables present some modern data for Venus in a way that aims to separate that planet's two anomalies. Familiarity with these data, and the phenomena they record, will also prove useful for understanding Ptolemy's treatment of Venus in Book X.

Table 1. Positions and Elongations of Venus, Jan 2010–May 2012

Date	Venus	Mean Sun	Elonga-tion	Date	Venus	Mean Sun	Elonga-tion
2010				Jun 20	126°34′	88°06′	38°28′E
Jan 1	277°51′	280°32′	2°42′W	Jun 25	132 21	93 02	39 20
Jan 6	284 08	285 28	1 20	Jun 30	138 06	97 57	40 09
Jan 11	290 26	290 24	0 02	Jul 5	143 48	102 53	40 55
Jan 16	296 43	295 19	1 24 E	Jul 10	149 27	107 49	41 39
Jan 21	303 00	300 15	2 45	Jul 15	155 03	112 44	42 19
Jan 26	309 17	305 11	4 06	Jul 20	160 35	117 40	42 55
Jan 31	315 34	310 06	5 27	Jul 25	166 03	122 36	43 27
Feb 5	321 50	315 02	6 48	Jul 30	171 25	127 31	43 54
Feb 10	328 06	319 58	8 09	Aug 4	176 43	132 27	44 15
Feb 15	334 22	324 54	9 29	Aug 9	181 53	137 23	44 31
Feb 20	340 38	329 49	10 48	Aug 14	186 57	142 19	44 38
Feb 25	346 53	334 45	12 08	Aug 19	191 52	147 14	44 38
Mar 2	353 07	339 41	13 26	Aug 24	196 37	152 10	44 27
Mar 7	359 21	344 36	14 44	Aug 29	201 09	157 06	44 04
Mar 12	5 34	349 32	16 02	Sep 3	205 28	162 01	43 27
Mar 17	11 46	354 28	17 19	Sep 8	209 29	166 57	42 32
Mar 22	17 58	359 23	18 35	Sep 13	213 10	171 53	41 18
Mar 27	24 09	4 19	19 50	Sep 18	216 26	176 48	39 38
Apr 1	30 20	9 15	21 05	Sep 23	219 11	181 44	37 27
Apr 6	36 29	14 10	22 18	Sep 28	221 19	186 40	34 39
Apr 11	42 37	19 06	23 31	Oct 3	222 42	191 35	31 07
Apr 16	48 45	24 02	24 43	Oct 8	223 14	196 31	26 42
Apr 21	54 52	28 58	25 54	Oct 13	222 47	201 27	21 20
Apr 26	60 57	33 53	27 04	Oct 18	221 21	206 23	14 59
May 1	67 02	38 49	28 13	Oct 23	219 02	211 18	7 44
May 6	73 05	43 45	29 20	Oct 28	216 08	216 14	0 05 W
May 11	79 07	48 40	30 27	Nov 2	213 08	221 10	8 02
May 16	85 09	53 36	31 32	Nov 7	210 30	226 05	15 36
May 21	91 08	58 32	32 37	Nov 12	208 37	231 01	22 23
May 26	97 07	63 27	33 39	Nov 17	207 44	235 57	28 13
May 31	103 04	68 23	34 40	Nov 22	207 51	240 52	33 02
Jun 5	108 59	73 19	35 40	Nov 27	208 56	245 48	36 53
Jun 10	114 52	78 15	36 38	Dec 2	210 51	250 44	39 53
Jun 15	120 44	83 10	37 34	Dec 7	213 29	255 39	42 11

The table lists longitudes (degrees eastward along the zodiac) of Venus and the mean sun at five-day intervals over a 29-month period; values are rounded to the nearest minute. Notice the following events: GEE about Aug 17, 2010, station about Oct 8, 2010 and again about Nov 20, 2010, and GWE about Jan 6, 2011. Another GEE occurs about Mar 26, 2012.

Date	Venus	Mean Sun	Elonga-tion
Dec 12	216°41′	260°35′	43°54′W
Dec 17	220 23	265 31	45 08
Dec 22	224 28	270 27	45 59
Dec 27	228 52	275 22	46 30
2011			
Jan 1	233 34	280 18	46 44
Jan 6	238 28	285 14	46 46
Jan 11	243 33	290 09	46 36
Jan 16	248 47	295 05	46 18
Jan 21	254 09	300 01	45 51
Jan 26	259 38	304 56	45 18
Jan 31	265 12	309 52	44 40
Feb 5	270 51	314 48	43 57
Feb 10	276 33	319 44	43 10
Feb 15	282 19	324 39	42 20
Feb 20	288 08	329 35	41 27
Feb 25	293 59	334 31	40 32
Mar 2	299 52	339 26	39 34
Mar 7	305 47	344 22	38 35
Mar 12	311 43	349 18	37 35
Mar 17	317 41	354 13	36 33
Mar 22	323 39	359 09	35 30
Mar 27	329 39	4 05	34 26
Apr 1	335 40	9 00	33 21
Apr 6	341 41	13 56	32 15
Apr 11	347 43	18 52	31 09
Apr 16	353 45	23 48	30 03
Apr 21	359 47	28 43	28 56
Apr 26	5 50	33 39	27 49
May 1	11 54	38 35	26 41
May 6	17 57	43 30	25 33
May 11	24 01	48 26	24 25
May 16	30 05	53 22	23 17
May 21	36 09	58 17	22 08
May 26	42 14	63 13	20 59
May 31	48 19	68 09	19 50
Jun 5	54 24	73 04	18 41
Jun 10	60 29	78 00	17 31
Jun 15	66 34	82 56	16 21

Date	Venus	Mean Sun	Elonga-tion
Jun 20	72°40′	87°52′	15°11′W
Jun 25	78 46	92 47	14 01
Jun 30	84 53	97 43	12 50
Jul 5	91 00	102 39	11 39
Jul 10	97 07	107 34	10 27
Jul 15	103 15	112 30	9 15
Jul 20	109 23	117 26	8 02
Jul 25	115 32	122 21	6 49
Jul 30	121 41	127 17	5 36
Aug 4	127 51	132 13	4 22
Aug 9	134 01	137 09	3 07
Aug 14	140 11	142 04	1 53
Aug 19	146 23	147 00	0 37
Aug 24	152 34	151 56	0 38 E
Aug 29	158 46	156 51	1 54
Sep 3	164 58	161 47	3 11
Sep 8	171 10	166 43	4 27
Sep 13	177 22	171 38	5 44
Sep 18	183 35	176 34	7 01
Sep 23	189 48	181 30	8 18
Sep 28	196 01	186 25	9 35
Oct 3	202 14	191 21	10 53
Oct 8	208 27	196 17	12 10
Oct 13	214 40	201 13	13 27
Oct 18	220 53	206 08	14 45
Oct 23	227 06	211 04	16 02
Oct 28	233 19	216 00	17 20
Nov 2	239 32	220 55	18 37
Nov 7	245 45	225 51	19 54
Nov 12	251 58	230 47	21 11
Nov 17	258 10	235 42	22 28
2011			
Nov 22	264 23	240 38	23 45
Nov 27	270 35	245 34	25 01
Dec 2	276 47	250 29	26 18
Dec 7	282 59	255 25	27 34
Dec 12	289 10	260 21	28 49
Dec 17	295 21	265 17	30 04
Dec 22	301 31	270 12	31 18

Date	Venus	Mean Sun	Elongation
Dec 27	307°40′	275°08′	32°32′ E
2012			
Jan 1	313 49	280 04	33 45
Jan 6	319 57	284 59	34 57
Jan 11	326 03	289 55	36 08
Jan 16	332 07	294 51	37 17
Jan 21	338 11	299 46	38 25
Jan 26	344 13	304 42	39 31
Jan 31	350 12	309 38	40 34
Feb 5	356 09	314 34	41 36
Feb 10	2 03	319 29	42 34
Feb 15	7 54	324 25	43 29
Feb 20	13 41	329 21	44 21
Feb 25	19 25	334 16	45 09
Mar 1	25 04	339 12	45 52

Date	Venus	Mean Sun	Elongation
Mar 6	30°37′	344°08′	46°29′ E
Mar 11	36 04	349 03	47 01
Mar 16	41 25	353 59	47 26
Mar 21	46 38	358 55	47 43
Mar 26	51 41	3 50	47 51
Mar 31	56 34	8 46	47 48
Apr 5	61 15	13 42	47 33
Apr 10	65 40	18 38	47 02
Apr 15	69 47	23 33	46 14
Apr 20	73 34	28 29	45 05
Apr 25	76 55	33 25	43 30
Apr 30	79 44	38 20	41 24
May 5	81 56	43 16	38 40
May 10	83 24	48 12	35 12

6. *Station and retrogradation.* In the preceding table, notice Venus' stations on about Oct 8 and Nov 20, 2010 as well as its retrograde motion between those stations.

7. *Greatest, least, and mean notion.* As Ptolemy will note in Chapter 3 below, the motion of Venus (averaged over a sufficiently long time) is equal to the motion of the mean sun. On a daily basis, therefore, the planet's mean motion must be equal to the motion of the mean sun, about 0° 59′ per day.

Notice in the preceding table that when the planet's elongation is zero, it is either overtaking the mean sun as on Jan 11, 2010, or falling behind it as on Oct 28, 2010. Examining the changes in Venus's position over the intervals before and after these dates reveals that the planet exhibits its *greatest eastward motion* as it overtakes the mean sun and its *least eastward motion* (or, rather, its greatest retrogradation) as it falls behind. These episodes correspond to what we earlier identified as greatest and least motions, respectively, of the apparent sun.

Furthermore, an occasion of GEE or GWE is also an occasion of *mean motion*—since when the planet exhibits a greatest elongation it must have ceased increasing its distance from the mean sun, but is not yet decreasing it. If its distance from the mean sun is neither increasing nor decreasing, the planet must be moving at the same rate as the mean sun. But the motion of the mean sun is, as we have noted, equal to the mean motion of the planet.

Observe, then, in the preceding table, that the planet's time from greatest to mean motion is from Jan 11 to Aug 17, 2010 or 218 days; while its time from mean to least motion is from Aug 17 to Oct 28, 1010 or 72 days. Thus the planet's time from greatest to mean motion is greater than its time from mean to least motion. Ptolemy will draw a conclusion from this inequality in Chapter 5 below.

8. *Magnitude of greatest elongation.* In the preceding table, notice that the GEE of Aug 17, 2010 (43°38′) does not equal the GEE of Mar 26, 2012 (47°51′). In fact, a survey of greatest elongations attained over a sufficient range of years reveals that both GEE and GWE vary between minimum and maximum values. The following table summarizes one such survey.

Table 2. Successive GEE and GWE of Venus, 1900–1939

Date	Mean Sun	GEE	Date	Mean Sun	GWE
1900 Apr 27	34°31′	47°14′	1900 Sep 19	177°26′	47°52′
1901 Dec 10	258 01	46 25	1902 Apr 28	35 02	44 26
1903 Jul 5	102 03	45 23	1903 Nov 25	242 46	47 55
1905 Feb 18	327 17	48 02	1905 Jul 12	109 13	45 48
1906 Sep 19	176 59	44 33	1907 Feb 6	314 59	45 39
1908 Apr 25	32 37	47 17	1908 Sep 17	175 32	47 50
1909 Dec 7	255 08	46 19	1910 Apr 25	32 07	44 26
1911 Jul 3	99 54	45 27	1911 Nov 23	240 51	47 58
1913 Feb 16	325 23	48 01	1913 Jul 9	106 20	45 43
1914 Sep 16	174 06	44 32	1915 Feb 3	312 05	45 44
1916 Apr 22	29 43	47 21	1916 Sep 15	173 37	47 46
1917 Dec 5	253 13	46 15	1918 Apr 23	30 13	44 26
1919 Jul 1	98 00	45 31	1919 Nov 20	237 57	48 00
1921 Feb 13	322 29	48 00	1921 Jul 7	104 25	45 39
1922 Sep 14	172 11	44 32	1923 Feb 1	310 11	45 49
1924 Apr 20	27 48	47 24	1924 Sep 13	171 73	47 43
1925 Dec 2	250 19	46 09	1926 Apr 20	27 20	44 25
1927 Jun 28	95 06	45 34	1927 Nov 18	236 03	48 03
1929 Feb 11	320 34	47 58	1929 Jul 4	101 31	45 35
1930 Sep 11	169 17	44 31	1931 Jan 29	307 17	45 54
1932 Apr 18	25 54	47 27	1932 Sep 10	168 49	47 40
1933 Nov 30	248 25	46 04	1934 Apr 18	25 25	44 26
1935 Jun 26	93 11	45 38	1935 Nov 16	234 08	48 05
1937 Feb 9	318 40	47 57	1937 Jul 2	99 37	45 31
1938 Sep 8	166 24	44 32	1939 Jan 27	305 22	45 59

In the previous table, dates are rounded to the nearest day at 00:00 UT; angles are rounded to the nearest minute. Notice that in any eight-year period beginning with a GEE, Venus attains GEE again five times, ending up with a nearly equal value of GEE on almost the same month and day at which it began. The same is true for GWE.

The values recorded in the previous table indicate that the magnitudes of GEE and GWE depend upon the month and day on which each greatest elongation is attained—or, what is the same, that they depend on the position of the mean sun. That dependency becomes manifest when we tabulate the greatest elongations in the order of the mean sun's longitude:

Table 3. Venus GEE and GWE by Longitude of Mean Sun

Date	Mean Sun	GEE	GWE
1934 Apr 18	25°25′		44°26′
1932 Apr 18	25 54	47°27′	
1926 Apr 20	27 20		44 25
1924 Apr 20	27 48	47 24	
1916 Apr 22	29 43	47 21	
1918 Apr 23	30 13		44 26
1910 Apr 25	32 08		44 26
1908 Apr 25	32 37	47 17	
1900 Apr 27	34 31	47 14	
1902 Apr 28	35 02		44 26
...			
1935 Jun 26	93 11	45 38	
1927 Jun 28	95 06	45 34	
1919 Jul 1	98 00	45 31	
1937 Jul 2	99 37		45 31
1911 Jul 3	99 44	45 27	
1929 Jul 4	101 31		45 35
1903 Jul 5	102 03	45 23	
1921 Jul 7	104 25		45 39
1913 Jul 9	106 20		45 43
1905 Jul 12	109 13		45 48
...			
1938 Sep 8	166 24	44 32	
1932 Sep 10	168 49		47 40
1930 Sep 11	169 17	44 31	
1924 Sep 13	171 73		47 43
1922 Sep 14	172 11	44 32	
1916 Sep 15	173 37		47 46
1914 Sep 16	174 06	44 32	
1908 Sep 17	175 32		47 50
1906 Sep 19	176 59	44 33	
1900 Sep 19	177 26		47 52
...			

Date	Mean Sun	GEE	GWE
1935 Nov 16	234°08′		48°05′
1927 Nov 18	236 03		48 03
1919 Nov 20	237 57		48 00
1911 Nov 23	240 51		47 58
1903 Nov 25	242 46		47 55
1933 Nov 30	248 25	46°04′	
1925 Dec 2	250 19	46 09	
1917 Dec 5	253 13	46 15	
1909 Dec 7	255 08	46 19	
1901 Dec 10	258 01	46 25	
...			
1939 Jan 27	305 22		45 59
1931 Jan 29	307 17		45 54
1923 Feb 1	310 11		45 49
1915 Feb 3	312 05		45 44
1907 Feb 6	314 59		45 39
1937 Feb 9	318 40	47 57	
1929 Feb 11	320 34	47 58	
1921 Feb 13	322 29	48 00	
1913 Feb 16	325 23	48 01	
1905 Feb 18	327 17	48 02	

It is evident that Table 3 lacks information for large portions of the zodiac; to fill these gaps would require more than an additional 150 years of data. But even with these limitations, the table suggests that for any single position of the mean sun in the zodiac, there is one characteristic value of GEE and one characteristic value of GWE. In Proposition 9.3 below, Ptolemy will show that occasions on which GWE and GEE are equal correspond to positions of the mean sun that are equidistant from the planet's apogee or perigee. In the above table, note that 1919 Jul 1 and 1937 Jul 2 are two such occasions. We will consider them more fully in the Preliminaries to Book X.

The Planets and the Mean Sun

We noted earlier that the mean motion of Venus is equal to that of the mean sun; the same, moreover, is true for Mercury. Such equality is a necessary consequence of these planets' limited elongations from the mean sun. For if the mean motions of sun and planet were not equal, then over time the planet would move further and further away from the mean sun, eventually exceeding any limit. Ptolemy does not explain why Mercury and Venus should have any connection whatever to the mean sun, still less why that relation should be the special relation of equality. He simply notes it as a feature that characterizes them.

In contrast, Mars, Jupiter, and Saturn each appear at all elongations from the mean sun. Necessarily, their mean motions do not equal the mean sun's and are in fact far slower. Yet these planets too exhibit a definite relation to the mean sun; for each of them, the sum of its longitudinal and its anomalistic motions proves to be equal to the motion of the mean sun over the same period of time, as Ptolemy will note in Chapter 3. Again, Ptolemy simply states this relation as a characteristic of the three planets.

Thus there are two characteristic relations between the planets and the mean sun:

$$s = l + a$$

for Mars, Jupiter, and Saturn, where s is the motion of the mean sun, l the mean motion in longitude, and a the mean motion in anomaly; and

$$s = l$$

for Mercury and Venus.

Book IX

1. On the Order of the Spheres of the Sun, Moon, and Five Planets.

H206.6 Now the preceding books {VII–VIII} would comprise more or less all that one might note by way of summary about the fixed stars, to the extent that appearances up to the present support progress in understanding. Since the treatment of the five planets remains for this work, we will fashion our exposition of them by combining each of our approaches, to the extent that their commonality permits, so as not to repeat ourselves.

Now, first we see that virtually all of the foremost mathematicians are in agreement about the order of their spheres, which in fact maintain their positions around the poles of the oblique, mid-zodiac circle: that they are all nearer to the earth than the sphere of the fixed stars, but farther from the earth than the sphere of the moon; and that three are farther from the earth H207 than the rest and the sphere of the sun (the sphere of the planet Saturn is larger, the sphere of the planet Jupiter is second in nearness to the earth, and the sphere of the planet Mars is beneath that). But we see that the sphere of the planet Venus and the sphere of the planet Mercury are placed below the sphere of the sun by the more ancient mathematicians, while they too are placed above it by some of the subsequent mathematicians, because the sun is never occulted[1] by them at all. We think that this judgement is unreliable; because some planets may be below the sun but not invariably also in one of the planes through the sun and our sight, but rather in a different plane. And, for this reason, they do not appear to occult it, just as occultations do not, for the most part, occur even when the moon passes between the earth and sun at conjunctions.[2]

[1] One might expect the sun to be "occulted" (hidden) by the planet if the planet were to come between the sun and the observer.

[2] Two celestial bodies are said to be in "conjunction" when they appear in the same degree of the zodiac, even if their latitude is not the same.

This notion cannot succeed in any other way, because none of the stars[1] makes any perceptible parallax[2] (the only appearance from which distances are obtained). Hence, the ordering of the more ancient mathematicians appears more persuasive. By placing the sun in the middle, their ordering separates in a more natural way the planets that stand at every distance away from the sun from those that do not, but are always carried about it;[3] to the extent, at any rate, that it does not displace the latter so far towards the earth that any perceptible parallax can be produced.

2. On the Aim of the Hypotheses of the Planets.

The preceding {IX.1}, then, would be our treatment of the ordering of the spheres. We propose to demonstrate that, just as for the sun and moon, all the apparent anomalistic motions of the five planets are produced through uniform, circular motions; these are proper to the nature of what is divine, but foreign to disorder and variability. Hence it is appropriate to regard success in such an aim as important and, in truth, as the goal of mathematical inquiry (θεωρία) in philosophy, but difficult for many reasons and something that, for good reason, no one has yet accomplished. For in the inquiries into the periodic motions of each planet what can be missed by sight in comparing observations, due to their subtlety, produces in the ensuing time-span a discrepancy that is perceptible sooner when one has made the examination over a shorter interval, but perceptible later when over a longer interval.[4] Hence, the time-span, beginning from which we possess recorded observations, being brief in comparison with a conception on so vast a scale ⟨of time⟩, renders prediction over a time-span that is many times H208.3

[1] Here, as often elsewhere, Ptolemy uses the words "stars" in a general way, not simply in reference to the "fixed" stars.

[2] Parallax is the difference in direction of a celestial object as seen by an observer from two positions separated by a known distance. This distance, together with the observed angle of directional difference, allows the distance of the body from the observer to be calculated trigonometrically. But since none of the planets exhibits perceptible parallax, as Ptolemy has just noted, he cannot use parallax to calculate the relative distances of the planets. The moon alone shows parallax; but Ptolemy's discussion of that topic (in Book V, chapters 11–13) is not included in this edition.

[3] For the reasons given here, Venus and Mercury are sometimes called the "inner" planets and the other three, the "outer" planets. Ptolemy does not use these terms.

[4] Ptolemy is describing the effects of observational error. If the period is determined using observations that are relatively near each other in time, a small error will be magnified when the period is projected into the future. A determination using more widely spaced observations will be proportionally less affected by the error, and will therefore have less of an effect on future projections.

larger unreliable. And in the inquiry into their anomalistic motions what produces no little confusion is, on the one hand, the fact that two anomalistic motions seem to arise for each planet.[1] These anomalistic motions are H209 unequal in their magnitudes and the durations of their periodic returns; one is observed to have a relationship to the sun, the other to the parts of the zodiac. Both are completely intermixed, so that what is distinctive in each is in consequence hard to distinguish. And, on the other hand, what produces no little confusion is the fact that the overwhelming majority of the ancient observations were inattentively as well as roughly recorded. For the more continuous observations among them contain stations and apparitions, and there is no apprehension of either of these distinctive features that is not subject to dispute. The stations cannot exhibit their true duration, since the change of place is imperceptible for many days both before and after the station itself. And the apparitions not only make the places,[2] along with what is seen first or last, immediately disappear, but also can be mistaken as to their durations on account of the difference in climates and sight of the observers.[3] And, in general, observations made in relation to one of the fixed stars from a rather large ⟨angular⟩ distance involve a magnitude of measurement that is difficult to calculate and subject to guesswork, unless one above all pays attention to them with scientific clarity (*διορατικῶς τε καὶ ἐπιστημονικῶς*). This difficulty arises not only because the lines between the stars that are being observed make angles that are different and not invariably right relative to the mid-zodiac circle; hence it is probable that a great error H210 results because of the variability of the inclination of the mid-zodiac circle in determining the longitudinal and latitudinal position. But this difficulty arises also because the same distances appear greater to the eyes at horizons but smaller at culminations; and, for this reason, they evidently can be measured sometimes as greater, at other times smaller, than the distance occurring in reality.

Hence I think that for all these reasons and, above all, because he had not received from earlier times the wealth of accurate observations that he himself provided to us, Hipparchus, a lover of truth in the highest degree, did investigate the hypotheses of the sun and the moon and, as far as was possible,

---

[1] *two anomalistic motions seem to arise for each planet*: See the discussion of Planetary Anomalies in the Preliminaries to Book IX.

[2] *the places*: that is, the places of the planets in the zodiac.

[3] *the difference in climates and sight of the observers*: The planets disappear as they approach the sun, then reappear as they depart from it. In either case, a planet and its position can only be seen for a short time; and its visibility will depend on the observer's location and atmospheric conditions.

prove by every means that they are produced through uniform, circular motions; but he did not contribute even a starting point for the hypotheses of the five planets, to judge by the commentaries, at any rate, that have reached us; he only arranged the observations of them in a quite useful way and proved through them that the appearances were inconsistent with the hypotheses of the mathematicians of that time. For he thought it necessary, as it seems, not only to prove that each planet makes a double anomalistic motion, or that the retrogradations are unequal and of such and such a size for each planet (for the rest of the mathematicians made geometrical demonstrations about one and the same anomalistic motion and one and the same retrogradation); or even that it happens that these anomalistic motions and retrogradations are produced either through eccentric circles or through circles concentric with the zodiac circle but carrying epicycles, or even, by Zeus, by both combined (the anomalistic motion relative to the zodiac circle being of such and such a size, the anomalistic motion relative to the sun of such and such a size). For all those who wished to prove uniform, circular motion through the so-called "eternal table"[1] by and large directed their attention to these matters. Yet they did so mistakenly and without proof as well; some did not follow through on the task at all, others to a certain extent only. He reasoned that it will not suffice for one who had progressed through all the mathematical sciences to such a degree of precision and love of truth to stop short with so little, as if he were no different from the rest. But, he reasoned, it would be necessary for the one who would persuade himself and future readers to demonstrate the magnitude of each of the two anomalistic motions and their periods through appearances that are clear and agreed upon. And it would be necessary for him, again, by mixing both the position and order of the circles, through which the anomalistic motions arise, to discover both the form of their motion, and then to harmonize virtually all the appearances with the specific form of the hypotheses of the circles. This task, I suppose, seemed difficult to him.

H211

We have stated these things not for the sake of show, but so that we acquiesce if we are compelled in some places by the subject-matter itself: either to employ something contrary to reason (*παρὰ τὸν λόγον*), as when, let us say, we make demonstrations through what is easy to follow with simple circles that are described in their spheres by their motion and are in the same plane as the mid-zodiac circle;[2] or to assume some things as elements not on the basis of a manifest first principle, but rather that are apprehended by

H212

[1] *eternal table*: ἡ καλουμένη αἰώνιος κανονοποιία. Surviving ancient texts make few references to tables by this name, and scholars disagree about how they were organized.

[2] *circles ... in the same plane as the mid-zodiac circle*: Ptolemy will set aside consideration of the planets' latitudinal deviations from the plane of the mid-zodiac circle until Book XIII.

continuous trial and adjustment;[1] or not to assume in all cases the same, invariant mode of motion or inclination of the circles.[2] For we know that the employment of any of these sorts of things will not at all impair our task, insofar as no significant difference will follow from it. And we know that things assumed without proof, once they are found in agreement with appearances, cannot have been discovered without some methodology and care, even if it is difficult to set forth how they are apprehended; since, generally speaking, the cause of the first principles is either nothing or what is by nature difficult to interpret. And we know that no one would consider it surprising and illogical that the mode of the hypothesis of circles differs in some places, as even the appearances of the stars themselves are found to be variable; for along with the preservation of uniform, circular motion in all cases without exception, all of the appearances as well are demonstrated according to the dominant, over-arching correspondence among the hypotheses.

H213 We have, accordingly, employed for individual demonstrations those observations that most admit of being undisputed; that is to say, those that have been carefully observed in an apparent contact or great approach of stars or even the moon, and above all those obtained through astrolabe instruments.[3] For our sight, directed as it were by the diametrically opposite openings in the rings, sees equal distances in every direction through similar arcs; and our sight is capable of precisely perceiving the longitudinal and latitudinal passages of each planet relative to the mid-zodiac circle, by rotation of both the ring in the astrolabe, corresponding to the zodiac circle, and the diametrically opposite openings, that correspond to the circles through its poles, towards what is being observed.

3. On the Periodic Returns of the Five Planets.

Since, then, these things have been distinguished beforehand in this way, we will set out first the periodic, that is to say, the shortest joint returns[4] of

---

[1] *to assume some things ... by continuous trial and adjustment*: Ptolemy will find it necessary, for example, to treat the planets' deferent circles as eccentric to the earth; this eccentricity is not founded on any first principle but is demanded by the particulars of planetary observations. Another example is his use of an additional circle (called the "equant") to account for planetary appearances. This circle will be formally introduced in Book IX, Chapter 6; it will be employed for particular planets in Book X.

[2] *not to assume in all cases...*: Mercury's appearances, in particular, are more complex than those of the other four planets and will require a unique configuration of hypotheses.

[3] *astrolabe instruments*: Ptolemy describes the astrolabe in detail in Book V, Chapter 1.

[4] *joint returns*: A *joint return* (συναποκατάστασις) occurs when the planet returns to the same position in the zodiac, and with the same relation to the mean sun, as it had at an earlier time.

each of the five planets as calculated with the greatest precision by Hipparchus. While a correction has been hit upon by us on the basis of comparison of the positions that came to light after the proofs of the anomalies, as we will make clear there, the joint returns are placed first by us in order to have readily set out, for calculations of the anomalies, the particular mean movements, of longitude and anomaly, of each planet; since there will be no significant difference here, even if someone employs the mean passages in a rather rough way. And in general by "motion of longitude" one must understand the motion of the center of the epicycle around its eccentric circle,[1] and by "anomaly" the motion of the star around its epicycle.[2]

H214

Calculations 9.1–9.5

We find, then, that 57 anomalies of the star Saturn fit exactly in 59 of our solar years (that is to say those from tropics or equinoxes to the same) and, in addition, $1\frac{3}{4}$ days, most nearly, and in two revolutions of the star and $1\frac{43}{60}$ degrees;[3] since, for the three stars that are always overtaken by the sun,[4] the sun always traverses, in the periodic return time of each, as many orbits as there are longitudinal revolutions of the star and periodic returns of anomaly added together.[5] And we find that 65 anomalies of the star Jupiter fit exactly in 71 solar years (similarly taken) less $4\frac{9}{10}$ days, most nearly, and in 6 revolutions of the star, those from tropics to the same tropics, less $4\frac{3}{4}$ degrees. And we find that 37 anomalies of the star Mars fit exactly in 79 of our solar years and $3\frac{13}{60}$ days, most nearly, and in 42 revolutions of the star (those from tropics to the same tropics) and $3\frac{1}{6}$ degrees. And we find that 5 anomalies of the star Venus fit exactly in 8 of our solar years, less $2\frac{3}{10}$ days, most nearly, and in 8 revolutions of the star, equal in number to those of the sun,[6] lacking $2\frac{1}{4}$ degrees. And we find that 145 anomalies of the star Mercury fit exactly in the 46 years and $1\frac{1}{30}$ days, most nearly, and, again, in an equal number of revolutions as the sun, 46, and 1 degree.

H215

---

[1] *the motion of the center of the epicycle around its eccentric circle*: that is, around its deferent, which is to be assumed eccentric to the earth. Ptolemy will determine the eccentricity of Venus' and Mars' deferents in Book X.

[2] *the motion of the star around its epicycle*: This motion must be figured with respect to the *mean apogee*, a specific point on the epicycle that Ptolemy will introduce in Chapter 6.

[3] The Table of the Sun's Mean Motion in Book III, Chapter 2 shows that $1\frac{3}{4}$ days corresponds to $1\frac{43}{60}$ degrees of longitudinal motion.

[4] *the three stars that are always overtaken by the sun*: that is, Saturn, Jupiter and Mars, whose mean longitudinal motions are slower than the sun's.

[5] *the sun always traverses ... as many orbits as there are longitudinal revolutions of the star and periodic returns of anomaly added together*: This is the relation $s = l + a$, discussed in the Preliminaries to Book IX.

[6] *8 revolutions of the star, equal in number to those of the sun*: This is the relation $s = l$, discussed in the Preliminaries to Book IX. In the next sentence Ptolemy acknowledges that Mercury exhibits the same relation.

But if we reduce for each planet the time-span of its periodic return to days, consistently with the annual period demonstrated by us {III.1}, and the number of its anomalies to the 360 degrees in one circle, we will have,

Calculations 9.6–9.10

for the star Saturn: 21,551;18 days and 20,520 degrees of anomaly;
H216
for the star Jupiter: 25,927;37 days and 23,400[1] degrees of anomaly;
for the star Mars: 28,857;53 days and 13,320 degrees of anomaly;
for the star Venus: 2,919;40 days and 1,800 degrees of anomaly;
and for the star Mercury: 16,802;24 days and 52,200 degrees of anomaly.

Accordingly, once we have divided, properly for each, the number of degrees of anomaly by the number of days, we will have as daily mean movement of anomaly,

Calculations 9.11–9.15

of Saturn: 0;57,7,43,41,43,40 degrees, most nearly;
of Jupiter: 0;54,9,2,46,26,0 degrees;
of Mars: 0;27,41,40,19,20,58 degrees;
of Venus: 0;36,59,25,53,11,28 degrees;
and of Mercury: 3;6,24,6,59,35,50 degrees.

H217
By taking $\frac{1}{24}$ of these for each planet, we will have as hourly mean movement of anomaly,

Calculations 9.16–9.20

of Saturn: 0;2,22,49,19,14,19,10 degrees;
of Jupiter: 0;2,15,22,36,56,5 degrees;
of Mars: 0;1,9,14,10,48,22,25 degrees;
of Venus: 0;1,32,28,34,42,58,40 degrees;
and of Mercury: 0;7,46,0,17,28,59,35 degrees.

Again, by multiplying the daily mean movements of anomaly of each planet by 30, we will have as monthly mean movement of anomaly,

Calculations 9.21–9.25

of Saturn: 28;33,51,50,51,50,0 degrees;
of Jupiter: 27;4,31,23,13,0,0 degrees;
of Mars: 13;50,50,9,40,29,0 degrees;
of Venus: 18;29,42,56,35,44,0 degrees;
and of Mercury: 93;12,3,29,47,55,0 degrees.

Similarly, by multiplying the daily mean movements of anomaly by the 365 days of one Egyptian year, we will have as annual mean movement of anomaly,

Calculations 9.26–9.30

of Saturn: 347;32,0,48,50,38,20 degrees;
of Jupiter: 329;25,1,52,28,10,0 degrees;
of Mars: 168;28,30,17,42,32,50 degrees;
of Venus: 225;1,32,28,34,39,15 degrees;
and of Mercury: 53;56,42,32,32,59,10 degrees of surplus.

[1] Here (II.216.1) Heiberg's text has 23,700.

Calculations 9.31–9.35

In like manner, by multiplying each of the annual mean movements of anomaly also by 18, we will have, just as for the table of days, as mean surplus of anomaly of an 18-year Egyptian period, H218

of Saturn:	135;36,14,39,11,30,0 degrees;
of Jupiter:	169;30,33,44,27,0,0 degrees;
of Mars:	152;33,5,18,45,51,0 degrees;
of Venus:	90;27,44,34,23,46,30 degrees;
and of Mercury:	251;0,45,45,53,45,0 degrees.

Calculations 9.36–9.50

The mean longitudinal movements also follow closely from these, so that we need not, by reducing the number of revolutions to degrees, also divide them by the time set out for each. Of the star Venus and the star Mercury it is clear that we will have the same mean movements as those already set out for the sun, while the mean movements of the remaining three stars fall short of them by the mean movements of anomaly to fill up the solar numbers, properly for each. And thereby we will have for the daily mean longitudinal movement,

of Saturn:	0;2,0,33,31,28,51 degrees;
of Jupiter:	0;4,59,14,26,46,31 degrees;
and of Mars:	0;31,26,36,53,51,33 degrees.

For the hourly movement,

of Saturn:	0;0,5,1,23,48,42,7,30 degrees;
of Jupiter:	0;0,12,28,6,6,56,17,30 degrees;
and of Mars:	0;1,18,36,32,14,39 degrees.

H219

For the monthly movement,

of Saturn:	1;0,16,45,44,25,30 degrees;
of Jupiter:	2;29,37,13,23,15,30 degrees;
and of Mars:	15;43,18,26,55,46,30 degrees.

For the annual movement,

of Saturn:	12;13,23,56,30,30,15 degrees;
of Jupiter:	30;20,22,52,52,38,35 degrees;
and of Mars:	191;16,54,27,38,35,45 degrees.

For the 18-years' mean motion,

of Saturn:	220;1,10,57,9,4,30 degrees;
the surplus of Jupiter:	186;6,51,51,53,34,30 degrees;
and the surplus of Mars:	203;4,20,17,34,43,30 degrees.

Accordingly, we will arrange again, for the sake of ease of usage, tables of the addition of the preceding mean movements for each of the stars in order, by 45 rows, similar to the rest, and 3 sections. Of these sections, the first will contain additions of eighteen-year periods, the second yearly and hourly additions, and the third the monthly and daily additions. And the tables follow.

[Chapter 4, "Tables of the mean motions of longitude and anomaly of the five planets," is omitted.]

5. Preliminaries to the Hypotheses of the Five Planets.

H250.3 Since the account of the anomalies that occur in the longitudinal passage of the five planets follows the setting out of these tables, we have gotten the rough conception of their outlines through the following considerations.

For the motions that are simplest and at the same sufficient for what is being proposed are two, as we said {H210.25–211.1}; one is produced by circles eccentric relative to the zodiac circle, and the other by circles that are concentric but carry epicycles. Similarly, the apparent anomalies for each individual star are also two; one is observed against the parts of zodiac circle, and the other against its configurations relative to the sun.[1] We find, then, in the latter anomaly, on the basis of frequent, different configurations observed around the same parts of the zodiac circle, that the time from greatest motion to mean motion always is greater than the time from mean motion to least motion for the five planets. This property {III.1, H I.208.11–14} cannot hold good on the eccentrical hypothesis, but rather its opposite, because the greatest passage always is produced at perigee on it, and on both hypotheses H251 the arc from perigee up to the point at mean passage is smaller than the arc from this up to apogee. But it can result on the hypothesis of epicycles, when, however, the greatest passage is produced not at perigee (just as for the moon) but at apogee; that is to say, when the star, beginning from its apogee,[2] makes its movement not westwards in the universe just like the moon, but rather eastwards. Hence we assume that such an anomaly results due to epicycles.

And for the anomaly observed relative to the parts of the zodiac circle we find, through arcs of the zodiac circle that pertain to the same phases[3] or the same configurations, that, in the opposite way, the time from least motion to

---

1 *configurations relative to the sun*: that is, elongations of the planet from the mean sun in the case of Mercury and Venus, or conjunctions and oppositions in the case of Mars, Jupiter, and Saturn.

2 *beginning from its apogee*: that is, its apogee on the epicycle. In the case of the planets it is necessary to distinguish apogee of the star on the epicycle from apogee of the center of the epicycle on its eccentric deferent.

3 *phases*: The Greek term φάσις has been translated elsewhere as "apparition." It can also mean simply "appearance."

mean motion is always greater than the time from mean motion to greatest motion. Such a property, again, can hold good on either hypothesis, in the manner we explained in our remarks on their similarity in the beginning of the treatise of the sun {III.3, H I.220.19–221.2}. But it is more suitable (*οἰκεῖον*) to the eccentrical hypothesis (according to which we assume that this sort of anomaly is produced) because the other anomaly has been found to be, as it were, unique to the epicyclic hypothesis alone.

Now, by applying particular observed passages to methods fashioned from the mixture of both hypotheses and continuous examination, we find that neither of the following assertions can be viable simply:[1] that the planes, in which we describe the eccentric circles, are unmoved, with the straight line through both their centers and the center of the mid-zodiac circle remaining always at the same distances from the tropic or equinoctial points (both apogees and perigees are observed along this line[2]); or that the epicycles have their own centers carried upon these eccentric circles, whose centers are those on which, when uniformly revolved in their eastward motion, they intercept equal angles in equal times. But, rather, we find that the apogees of the eccentric circles also perform some slight movement eastwards of the tropic points; this, again, is uniform around the center of the zodiac circle and for each star is, roughly speaking, that amount which the sphere of the fixed stars also has been found to be making; that is to say, 1 degree in 100 years,[3] so far as it is possible to detect at present. And we find that the centers of the epicycles are carried on circles that are equal to the eccentric circles that produce the anomaly,[4] but are not described with the same centers;[5]

H252

---

[1] *neither of the following assertions can be viable simply*: Ptolemy is about to name two reasonable suppositions which, nevertheless, cannot be maintained. The first supposition is that a planet's line of apsides is fixed; the second supposition is that the center of the planet's epicycle moves uniformly around the center of the eccentric circle.

[2] *this line*: that is, the line of apsides.

[3] *roughly ... 1 degree in 100 years*: Ptolemy attributes a slow eastward drift to each planet's line of apsides. Since he is able to ascertain only that this drift is "roughly" equal to the precession of the equinoxes (Book VII, Chapter 2), he cannot be sure whether each line of apsides participates in the equinoctial precession or possesses its own independent motion.

[4] *the eccentric circles that produce the anomaly*: Ptolemy will use this phrase again near the end of page H257 and will there make it clear that it refers to the circles about whose respective centers the planets execute uniform motion—that is, the *equant* circles.

[5] *not described with the same centers*: Ptolemy is anticipating features of his planetary hypotheses whose necessity will not be evident until Book X: Beginning with the following chapter, he will suppose that for each planet, the center of its epicycle is carried on an eccentric deferent which is equal to the equant circle but whose center differs both from the equant center and the center of the mid-zodiac circle.

but rather, for the rest of the planets,[1] they are described with centers that bisect the straight lines between the centers of those eccentric circles[2] and the center of the zodiac circle. We find that for the star of Mercury alone the ⟨eccentric⟩ circle is described with its center as far separated from the center that revolves it as both that center is separated from the center that produces the anomaly[3] relative to apogee and this center is separated from the center assumed for sight.[4] And, in fact, for this star alone also, just as for the moon as well, we find that the eccentric circle is also revolved by the stated center counter to its epicycle, again, westwards one revolution in a year; since it actually appears to be at nearest perigee twice in one revolution, just as the moon appears to be at nearest perigee twice in one month.

H253

6. On the Mode and Difference of the Hypotheses.

The mode of the hypotheses that are deduced through the foregoing might prove more easy to understand in the following way.

Proposition 9.1

For let there be conceived, on the hypothesis of the rest of the planets,[5] first, eccentric circle ABC around center D, and diameter ADC through D and the center of the zodiac circle. Upon this let the center of the zodiac circle E (that is to say the sight of observers) make A the point of farthest apogee, and C that of nearest perigee. After DE has been bisected at F, with center F and distance DA let circle GHI (equal, evidently, to circle ABC) be described, and with center H let epicycle JK be described, and let $JHKD$ be joined.[6]

H254

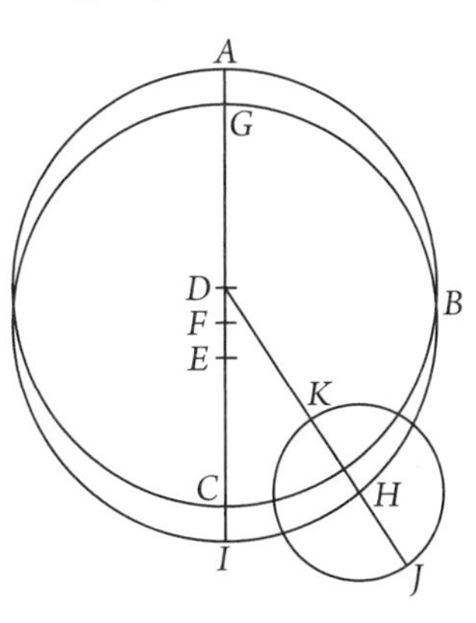

[1] *for the rest of the planets*: But not Mercury, as Ptolemy is about to state.

[2] *those eccentric circles*: that is, the aforementioned "eccentric circles that produce the anomaly"—the equant circles. Ptolemy will demonstrate in Book X that the equants of both Venus and Mars exhibit double the eccentricity of their corresponding eccentric deferents. Thus the centers of the eccentric deferents "bisect the straight lines" between the center of the zodiac and the respective equants. See the construction and diagram in Chap. 6 {H253}.

[3] *the center that produces the anomaly*: that is, the equant.

[4] For Mercury, Ptolemy will place the center of the eccentric on a small circle; thus there are four circles with the same spacing. We will not attend to the further details of Ptolemy's treatment of Mercury.

[5] *the rest of the planets*: that is, the planets other than Mercury.

[6] *with center F and distance DA let circle GHI ... be described*: Thus F is the center of the eccentric deferent GHI, while D is the center of the equant circle ABC.

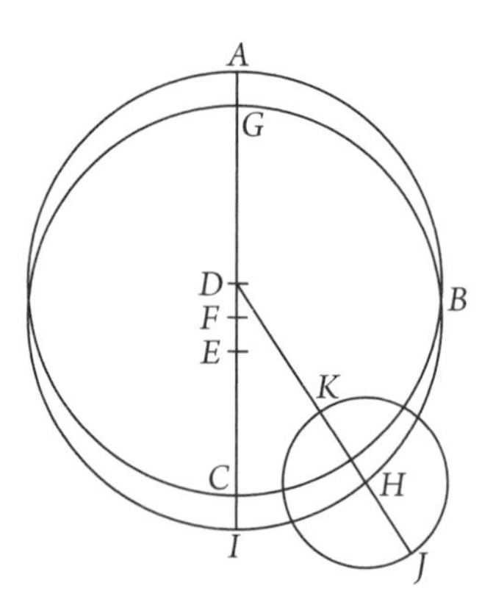

Now, we assume first that, given the latitudinal passage of the stars according to our later demonstrations {Book XIII}, the plane of the eccentric circles is oblique to the plane of the mid-zodiac circle and, furthermore, that the plane of the epicycle is oblique to the plane of the eccentric circles. But for the longitudinal passages we assume, for the sake of ready use, that all the circles are conceived in the single plane of the zodiac circle; since there will be no significant difference in longitude due to such small inclinations, at any rate, as will appear for each single star. Next, we say that the whole plane is uniformly revolved eastwards of the zodiac signs around center *E*, shifting the apogees and perigees[1] 1 degree in 100 years. And we say that diameter *JHK* of the epicycle is revolved by center *D*, again, uniformly eastwards of the zodiac signs, consistently with the periodic longitudinal return of the star, and that it rotates as well points *J* and *K* of the epicycle and center *H* that is always carried by eccentric circle *GHI*. And we say that the star itself also moves on epicycle *JK*, again, uniformly and relative to the diameter that always verges towards center *D*;[2] it makes its periodic returns consistently with the mean period of the solar anomaly,[3] and as the movement at apogee *J* is produced eastwards of the zodiac signs.[4]

H255

[Proposition 9.2 {H255.11–256.24}, on Mercury, is omitted.]

And we might explore still further what is being assumed through our later quantitative proofs for each individual planet. In the course of these proofs, those factors that have, somehow or other, moved us towards our conceptions of the hypotheses will also appear in a rather general way in many places.

H257

[1] *shifting the apogees and perigees*: Shifting them, that is, relative to the tropic and equinoctial points. As noted previously, Ptolemy regards the zodiac signs as fixed with respect to the tropics and equinoxes, while the constellations (and now, too, the lines of apsides) move eastwards.

[2] *uniformly and relative to the diameter that always verges towards center D*: The planet's motion on its epicycle is uniform with respect to line *JKD*, drawn through the center of the epicycle and the equant, the center of the epicycle's mean motion. The importance of point *J*, called the "mean apogee," will be evident in Book X, Chapter 6.

[3] *it makes its periodic returns consistently with the mean period of the solar anomaly*: The ratios of longitudinal to anomalistic periods offered in Chapter 3 above are invariant, provided that anomalistic motion is figured with respect to the mean apogee *J*.

[4] *as the movement at apogee J is produced eastwards*: That is, the planet's motion on the epicycle is in the same direction as the epicycle's motion on the deferent.

The longitudinal periods, however, do not coincide with points of the mid-zodiac circle and the apogees or perigees of the eccentric circles because of the assumed change of the latter.[1] One must premise, then, that the longitudinal motions set out by us in the preceding manner do not encompass periodic returns observed relative to the apogees of the eccentric circles; but rather, they encompass periodic returns that occur relative to the tropic and equinoctial points—in a manner consistent with our annual period.

Proposition 9.3

Now, it must first be shown that, on these hypotheses, when the mean, longitudinal passage of the star[2] is equally separated, on each side, from apogees or perigees, both the zodiacal anomalistic difference[3] and the greatest elongation on the epicycle in the same direction as the mean passage are equal at each distance.

For[4] let eccentric circle *ABCD*, on which the center of the epicycle is carried, be around center *E* and diameter *AEC*, upon which let *F* be assumed as center of the zodiac circle. And let the center of the eccentric circle that produces the anomaly, that is to say, around which we say that the mean passage of the epicycle is uniformly produced, be *G*.[5] And let *BGH* and *DGI* be drawn through, each equally separated from apogee *A*, so that angles *AGB* and *AGD* are equal. And let equal epicycles be described around points *B* and *D*, and let *BF* and *DF* be joined. And let straight lines *FJ* and *FK* be drawn from *F* (our sight) tangent to the epicycles in the same direction.[6]

H258

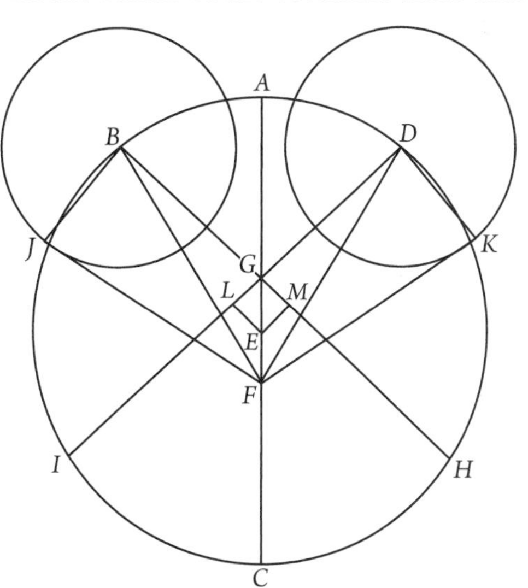

[1] *the assumed change of the latter*: that is, slow drift of each planet's apogee and perigee, which Ptolemy discussed in Chapter 5 {H252}.

[2] *the mean longitudinal passage of the star*: that is, the regular angular motion of the center of the epicycle measured around the equant.

[3] *the zodiacal anomalistic difference*: The difference in zodiacal position between where the center of the planet's epicycle will be (at a given time) if it moves regularly with respect to the equant, and where it would be if it moved regularly with respect to the earth.

[4] Heiberg's drawing {H258.5} is modified here.

[5] *let the center of the eccentric circle that produces the anomaly ... be G*: Thus *G* is the equant center.

[6] *in the same direction*: that is, either both towards or both away from apogee of the eccentric.

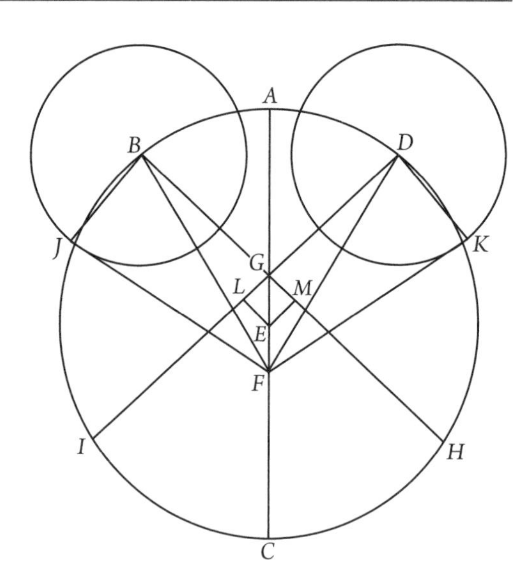

I say that angle *FBG* of the difference of the zodiacal anomaly is equal to angle *FDG*, and angle *BFJ* of the greatest elongation on the epicycle is, similarly, equal to angle *DFK*. For in this way, the magnitudes of elongations based on mixing greatest elongations from their mean passage[1] will also be equal.

Now, let *BJ* and *DK* be drawn perpendicular from *B* and *D* to *FJ* and *FK*, and *EL* and *EM* from *E* perpendicular to *BH* and *DI*.

Since angle *MGE* is equal to angle *LGE*,
and the angles at *L* and *M* are right,
and straight line *EG* of the equiangular triangles is common,
LG is equal to *MG*,
and perpendicular *EL* is equal to perpendicular *EM* [Euc. 1.26].

Therefore straight lines *BH* and *DI* are equally separated from center *E*; **H259**
therefore they themselves [Euc. 3.14] and their halves are equal.

So that also, the remainders *BG* and *DG* are equal.

But straight line *GF* also is common, and angle *BGF* contained by the respectively equal sides is equal to angle *DGF*;
and therefore base *BF* is equal to base *DF*,
and angle *GBF* is equal to angle *GDF* [Euc. 1.4].

Radius *BJ* of the epicycle also is equal to *DK*,
and the angles at *J* and *K* are right;
and therefore angle *BFJ* is equal to angle *DFK* [Euc. 1.4];
which it was proposed to show.[2]

[The remainder of Chapter 6 {H259.12–261.18}, on Mercury, is omitted.]

---

[1] *mixing greatest elongations from their mean passage*: that is, comparing GEEs with GWEs.

[2] Having proved that GEE and GWE are equal at longitudinal positions equidistant from apogee, Ptolemy can use this equality as a criterion by which to locate the line of apsides for Venus. He will do so in Book X, Chapter 1.

PRELIMINARIES TO BOOK X

Apogee and Perigee

Ptolemy deduced the criterion for locating Venus' line of apsides in Book IX, Chapter 6 (Proposition 9.3). There he proved that two positions of the mean sun, exhibiting GEE and GWE respectively equal, are equidistant from that line. Accordingly, Ptolemy's first task in Book X is to find the line of apsides for Venus; and in Chapter 1 he will determine that it extends through 25° Taurus and 25° Scorpio, that is, through longitudes 55° and 235°.

But which of these positions is apogee, which perigee? In Chapter 2 he notes that a greatest elongation—whether GEE or GWE—will be a maximum when the mean sun is at Venus' perigee, and a minimum when it is at Venus' apogee. In this diagram, *E* is our eye and *D* is the center of Venus' deferent. Thus *A* will be Venus' apogee, *C* its perigee; and it is evident that angle *GEC* (greatest elongation at perigee) will be larger than angle *FEA* (greatest elongation at apogee). Since Ptolemy finds Venus' greatest elongation to be smaller when the mean sun is at 25° Taurus than when the mean sun is at 25° Scorpio, he is able to locate Venus' apogee at 25° Taurus, that is, longitude 55°.

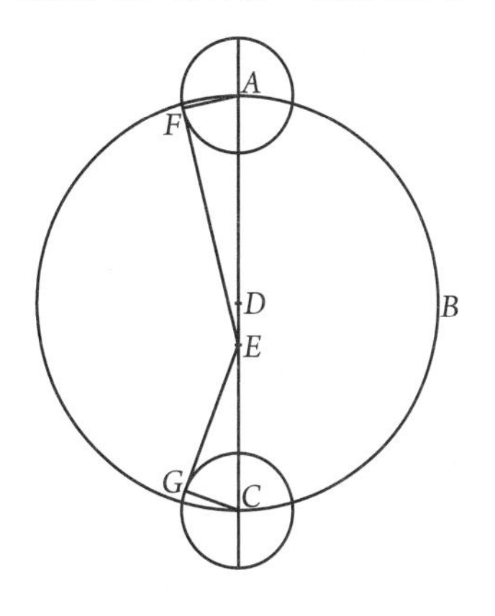

Applying the same method to modern data, we noted in the Preliminaries to Book IX that Venus' GEE and GWE were equal on 1919 Jul 1 (mean sun at longitude 98°00′) and 1937 Jul 2 (mean sun at longitude 99°37′) respectively; thus the modern apogee would appear to lie midway between these values, at 98°48′. If so, Venus' line of apsides will have moved eastward about 44 degrees in the roughly 18 centuries between Ptolemy's time and the modern data, some 2.4 degrees per century—a considerable discrepancy from Ptolemy's estimate of about 1 degree per century which he put forth in Book IX, Chapter 5. One might suppose that such a disparity merely represents an erroneous projection based on Ptolemy's too-limited range of data. In fact it points to real deficiencies in his planetary theory, deficiencies subsequently to be rectified by Kepler and Newton.

Eccentricity and the Equant

Ptolemy added the eccentric deferent and the equant to his planetary hypotheses in Chapters 5 and 6 of Book IX, without there explaining the considerations which necessitated such devices. In the later chapters of Book IX and in Book X he demonstrates them from selected planetary appearances—greatest elongations in the case of Mercury and Venus, oppositions in the cases of Mars, Jupiter, and Saturn. The elongations "make the epicycle visible," so to speak, and so reveal both the respective eccentricities of the planets' deferents and the respective locations of the planets' equants. Ptolemy's treatment of the other planets is less closely tied to their appearances, as we shall see.

In Chapter 2 below, Ptolemy will establish the eccentricity of Venus' deferent on the basis of its greatest elongations at apogee and perigee. Then in Chapter 3 he will determine the location of its equant from greatest elongations observed when the mean sun is "a quadrant's distance" (that is, a right angle) from apogee, as shown in this diagram from Chapter 3, in which *D* is Venus' equant center, *H* the center of its eccentric deferent, and *B* is our eye at the center of the zodiac circle. Line *BX* is drawn in the direction of the mean sun; and since the mean sun revolves about *B* with the same period as the mean planet revolves about equant *D*,[1] *BX* will always be parallel to *DE*.

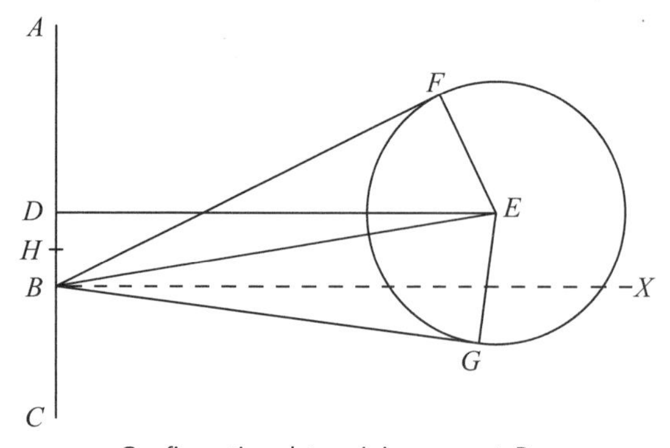

Configuration determining equant *D*

Ptolemy will find that Venus' equant center *D* lies on its line of apsides, with an eccentricity *BD* exactly twice the eccentricity *BH* of the planet's deferent—thus confirming his anticipatory remarks to that effect in Book IX.

When considering the other planets, however, Ptolemy does not offer independent demonstrations of their eccentricities. Instead, he determines their equant positions from observations, but assumes without discussion that for them too, just as for Venus, the eccentricities of their equants are double the eccentricities of their deferents, respectively.

---

[1] This is the relation $s = l$, discussed in the Preliminaries to Book IX.

Zodiacal Anomaly and the Two Eccentricities

Since Mars, Jupiter, and Saturn have unlimited elongations from the sun, the procedure used for Mercury and Venus cannot be applied to the rest of the planets. Instead, rather than "making the epicycle visible," as expressed in the preceding section, he makes it *invisible* by choosing select observations at times when the planet's observed position is aligned with the center of the epicycle: that is, when the planet is at true apogee or true perigee on the epicycle. Apogee, however, won't work, for the following interesting reason. As Ptolemy will prove in Prop. 10.7, because of the relation $s = l + a$ (see the Preliminaries to Book IX and Chapter 3 of that Book), the line from the center of the epicycle to the planet is always parallel to the line of sight from earth to the mean sun. As a consequence, when the planet is at apogee on the epicycle, beyond the epicycle's center, the line of sight to the sun must also be directed towards the epicycle's center; thus the planet will be aligned with the sun—and therefore invisible. In contrast, when the planet is at perigee on the epicycle, the line from the center of the epicycle to the planet is directed towards the earth, and therefore the line from the earth to the sun will be *opposite* the line from the earth to the planet. So, in effect, when one chooses observations at the time when the planet is opposite the mean sun in the sky, one is observing the position of the center of the epicycle, since it is aligned directly with the planet.

Near the beginning of Chapter 6 {H317}, Ptolemy states that the eccentricity of the equant (line *LH* in the diagram shown here) "is discovered through the greatest difference of zodiacal anomaly." One might suppose from this that he would choose several oppositions near this maximum difference. But in the absence of any direct index of this maximum (as, for example, the elongations, in the theory of Venus), its location is not readily obtainable, but must be found by comparing observed positions of the planet with the times of observations, at a number of locations around the circle.

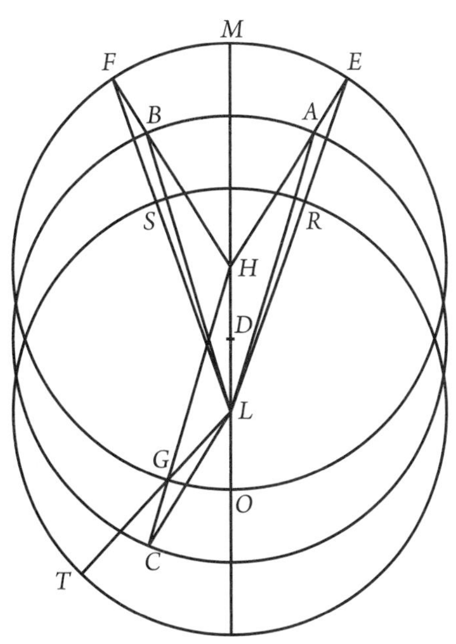

Ptolemy's strategy, therefore, will be to find, or make, three reliable observations of the planet at opposition to the mean sun and to determine

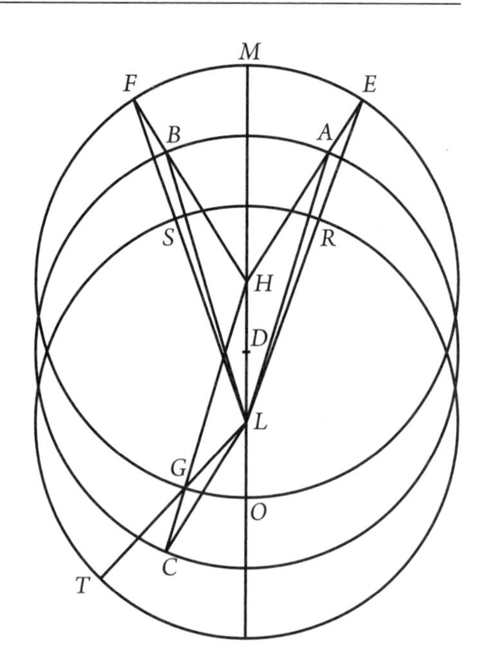

the eccentricity of the circle on which the center of the epicycle moves, *as if* that circle coincided with the circle around which the uniform motion takes place (the equant circle, *EFG*). This lengthy proof is presented in Props. 10.8 and 10.9. The observations are represented by the points *E*, *F*, and *G* in the adjacent diagram. The counterfactual supposition of a simple eccentric allows him to determine preliminary, trial values for the eccentricity *HL* of the equant *H* and the direction of the line of apsides *OM*, at the conclusion of Prop. 10.9. The arcs *EF* and *FG* are determined by the times of the observations, while the angles about *L* are given by the observed positions (Observations 10.12–14).

This model is perfectly correct for the three chosen observations, but, as Ptolemy must have known, it would not prove satisfactory at other positions. Accordingly, without any further reference to observations, he will suppose that the true eccentric circle bearing the center of the epicycle (the deferent) has its center at *D*, exactly halfway between the trial position of the center of the equant *H* and the center of the zodiac *L* (the earth). For the three chosen times, the center of the epicycle must still fall on the lines *HE*, *HF*, and *HG*, which represent the time-angles, and must therefore be at the points *A*, *B*, and *C*, respectively. These points must also remain at the same positions as observed from *L*. These requirements would appear to be contradictory: if the angles at *H* are to stay the same as before, it looks as if the angle *ALB* must be smaller than the previous angle *ELF* determined by the observations. Ptolemy's ingenious solution to this is to leave points *A*, *B*, and *C* fixed, as the observations require, and then to move point *H* downward slightly, thus adjusting the angles *EHF* and *FHG* to represent the times of observations correctly. This will also change the angles *ELF* and *FLG*, but *these angles are not determined by the observations,* and so may be allowed to change. The corrections to these angles (which he initially designates by the equivalent arcs *RS* and *ST* on the concentric circle) can be computed from the assumed position of the circle *ABC*. The computations of the trial corrections are taken up in Props. 10.10–14.

In practice, the way Ptolemy determines the new position of *H* is to apply these corrections to the angles *RLS* and *SLT* and then, on the basis of these new angles together with the unchanged arcs *EF* and *FG*, to go through the entire procedure again, using (as before) only the circle *EFG* and the new values for the angles about point *L*, whose position within the circle is unknown. This is stated, though not very clearly, in Prop. 10.15, where he writes, "...we followed the theorem proved previously [Props. 10.8–10.9], through which we demonstrate the apogee and ratio of eccentricity." He does not provide the details of the computation, "in order that we not make our explanation lengthy by repetition" {H339}, giving only the new eccentricity and position of perigee. This is the second pass through a truly remarkable iterative derivation of the parameters of the planetary model. Although this procedure can never provide *exact* positions of the circles and the apsides, it can approach the truth as nearly as is desired.

The difficulty that leads to the iterative method is this: any new position of point *H*, found by going through a new iteration of the proofs, results in new and slightly different corrections to the arcs *RS* and *ST*, which in turn require a further adjustment of the position of *H*. The third iteration will prove to be Ptolemy's final adjustment {H340}: fortunately for him, the determination becomes more accurate with successive iterations.

For convenience, here are the angles and arcs used by Ptolemy, in tabular form. First, the arcs and angles determined by the observations, which of course remain fixed throughout the demonstrations:

Arcs and angles obtained from observations (Calc. 10.14)		
Time arcs:	$\widehat{EF} = 81;44°$	$\widehat{FG} = 95;28°$
Observed angles:	$\measuredangle\, ILJ = 67;50°$	$\measuredangle\, JLK = 93;44°$

Note that initially $\measuredangle\, ILJ = \measuredangle\, ELF$ and $\measuredangle\, JLK = \measuredangle\, FLG$.

Next, the computed eccentricities and arcs:

Iteration	Eccentricity *HL*	Eccentricity *DL*	Arc *RS*	Arc *ST*
1	$13;7^P$ [Prop. 10.9 A]	$6;33\frac{1}{2}^P$ [Prop. 10.10]	68;55° [Prop. 10.12]	92;21° [Prop. 10.14]
2	$11;50^P$ [Prop. 10.15]	$5;55^P$ *	68;46° [H339]	92;36° [H339]
3	$12;0^P$ [H340]	$6;0^P$ *		

* These are not stated explicitly, but are taken as half of eccentricity *HL* (see, for example, H341 l. 1)..

Ptolemy concludes Chapter 7 by demonstrating that the planetary positions computed using the resulting model accurately reproduce the positions as found in the original observations. He also calculates Mars' position on the epicycle, which will be used in the following chapter.

Remarkably, this can be done even though the size of the epicycle has not yet been determined.

The Epicycle of Mars, At Last

It is only after Mars' eccentric circle is been determined that Ptolemy can turn his attention to the epicycle. To find its radius, he will make use of an observation taken a few days after the last of the three observations used in determining the eccentric circle. The observed position, together with the position of the center of the epicycle and Mars' angular position on the epicycle, will prove sufficient to establish the exact size of the epicycle (Prop. 10.20).

Book X

1. Demonstration of the Apogee of the Star Venus.

H296.3 The hypotheses, then, of the star Mercury and the sizes of its anomalies and, furthermore, the quantity of its periodic motions and their epochs were obtained by us in this way {IX.7–11, not included in the present selection}. For the star Venus, using the greatest elongations that are equal and in the same direction,[1] we again first inquired on what parts of the mid-zodiac circle the apogee and perigee of eccentricity are until we ran short of ancient observations that were in precise correspondence, and have fashioned the following idea from observations in our time.

Observation 10.1
Calculation 10.1

For among the observations given to us by Theon the mathematician, we found an observation recorded in the 16th year of Hadrian, by Egyptian reckoning Pharmouthi 21/22, in which he says that the evening star Venus was at greatest elongation from the sun, westwards of the middle of the Pleiades by the length of the Pleiades; and it appeared to have passed by them even slightly more southerly. Accordingly, since the middle of the Pleiades then occupied 3 degrees of Taurus, by our principles of reckoning, while its H297 length is $1\frac{1}{2}$ degrees, most nearly, the star Venus clearly occupied $1\frac{1}{2}$ degrees of Taurus then. So that, since the mean sun also occupied $14\frac{1}{4}$ degrees of Pisces then,[2] its greatest evening elongation from the sun's mean passage[3] was $47\frac{1}{4}$ degrees.

Observation 10.2
Calculation 10.2

We observed in the 14th year of Antoninus, by Egyptian reckoning Thoth 11/12, that the morning star Venus was at greatest elongation from the sun, and it was separated from the ⟨star in the⟩ middle knee of Gemini northwards and eastwards by half of one full moon.[4] And the fixed star, by our reckoning, then occupied $18\frac{1}{4}$ degrees of Gemini, so that the star Venus happened to be around $18\frac{1}{2}$ degrees, most nearly, while the mean sun occupied $5\frac{3}{4}$ degrees

---

[1] *in the same direction*: that is, either both toward apogee or both toward perigee of the eccentric, as in Book IX, Chapter 6, Proposition 9.3. Ptolemy is about to apply this proposition to determine Venus's apogee and perigee.

[2] The position of the mean sun may be determined for any given time using the Table of the Sun's Mean Motion (Book III, Chapter 2), starting from the sun's mean epoch (Book III, Chapter 7); or from the position of the visible sun, using the Table of the Sun's Anomaly (Book III, Chapter 6).

[3] *from the sun's mean passage*: that is, from the mean sun.

[4] *half of one full moon*: The full moon subtends 1/2 degree. Ptolemy notes this value in V.14, not included in this edition.

of Leo; therefore, its greatest morning elongation[1] also was $47\frac{1}{4}$ of the same degrees.[2]

Since according to the earlier observation its mean passage occupied $14\frac{1}{4}$ degrees of Pisces, and according to the second $5\frac{3}{4}$ degrees of Leo, and the point midway between them on the mid-zodiac circle falls out about 25 degrees of Taurus and Scorpio, the diameter of its apogee and perigee would be along these points.

Observation 10.3
Calculation 10.3

Among the observations from Theon we similarly found that, in the 12th year of Hadrian, by Egyptian reckoning Athur 21/22, the morning star Venus was at greatest elongation from the sun, eastwards of the star on the tip of the southern wing of Virgo, by the length of Pleiades or less by its own size; and it seemed to pass by the star more northerly by 1 moon. Since the fixed star then, by our reckoning, occupied $28\frac{11}{12}$ degrees of Leo, so that the star Venus also occupied $\frac{1}{3}$ degree, most nearly, of Virgo, while the mean sun occupied $17\frac{13}{15}$ degrees of Libra, its greatest morning elongation from the sun's mean passage was $47\frac{8}{15}$ degrees.

H298

Observation 10.4
Calculation 10.4

We observed in the 21st year of Hadrian, by Egyptian reckoning Mechir 9/10, in the evening, that the star Venus was at greatest elongation from the sun, and it was to the west of the most northerly of the four stars in the Quadrangle, beyond the eastern one, and in a straight line with the groin of Aquarius, $\frac{2}{3}$, most nearly, of a full moon and it seemed to illuminate (καταλάμπτειν) the star. So that, since again the fixed star then, by our reckoning, occupied 20 degrees of Aquarius, and, for this reason, the star Venus also was around $19\frac{3}{5}$ degrees, while the mean sun occupied $2\frac{1}{15}$ degrees of Capricorn, in this case also its greatest evening elongation was $47\frac{8}{15}$ of the same degrees. And the points of the mid-zodiac circle midway between $17\frac{13}{15}$ degrees of Libra (according to the first observation) and $2\frac{1}{15}$ degrees of Capricorn (according to the second observation) are about 25 degrees, most nearly, again of Scorpio and Taurus.

H299

2. On the Size of Its Epicycle.

The fact, then, that in our times the apogee and perigee of its eccentricity are about 25 degrees of Taurus and Scorpio was obtained by us through these calculations. Consequently, we in turn inquired into its greatest elongations from the sun's mean passage, when this passage occurs around 25 degrees of Taurus and around 25 degrees of Scorpio.

Observation 10.5
Calculation 10.5

For among the observations given to us by Theon we find, in the 13th year of Hadrian, by Egyptian reckoning Epiphi 2/3, that the morning star

---

[1] *greatest morning elongation*: that is, greatest western elongation.

[2] *the same degrees*: that is, degrees measured around the center of the mid-zodiac circle.

Venus was at greatest elongation from the sun, westwards of the straight line through the western star of the three in the head of Aries and the star on its hind leg by $1\frac{2}{5}$ degrees, and it made the distance to the western star of those in the head double, most nearly, the distance to the star on the leg. The western star of the three in the head of Aries then occupied, by our reckoning, $6\frac{3}{5}$ degrees of Aries and is more northerly than the mid-zodiac circle by $7\frac{1}{3}$ degrees; but the star in the hind leg of Aries occupied $14\frac{3}{4}$ degrees and is more southerly than the mid-zodiac circle by $5\frac{1}{4}$ degrees. Therefore, the star Venus occupied $10\frac{3}{5}$ degrees of Aries and was more southerly than the mid-zodiac circle by $1\frac{1}{2}$ degrees. So that, since the mean sun occupied then $25\frac{2}{5}$ degrees of Taurus, its greatest elongation from the sun's mean passage is $44\frac{4}{5}$ degrees.

H300

We observed, in the 21st year of Hadrian, by Egyptian reckoning Tubi 2/3, in the evening, that the star Venus was at greatest elongation from the sun, and, being sighted relative to the stars in the horns of Capricorn, appeared to occupy $12\frac{5}{6}$ degrees of Capricorn, while the mean sun occupied $25\frac{1}{2}$ degrees of Scorpio. So that in this case its greatest elongation from the sun's mean passage adds up to $47\frac{1}{3}$ degrees, and it has become clear that the apogee is about 25 degrees of Taurus, while the perigee about 25 degrees of Scorpio.

Observation 10.6
Calculation 10.6

It has become clear to us that the eccentric circle carrying the epicycle of the star Venus is stationary, because the sum of greatest elongations in each direction from the sun's mean passage is nowhere found on the mid-zodiac circle smaller than the sum in Taurus, nor larger than the sum in Scorpio.[1]

Proposition 10.1

H301

Proposition 10.2

Now, with these values assumed, let eccentric circle *ABC* (upon which the epicycle of the star Venus is always carried) be around diameter *AC*. On this let *D* be assumed as the center of the eccentric circle, *E* the center of the zodiac circle, and *A* the point near the 25th degree of Taurus. And let equal epicycles be described around points *A* and *C* (upon which are points *F* and *G*) and, when tangents *EF* and *EG* are drawn through, let *AF* and *CG* be joined.

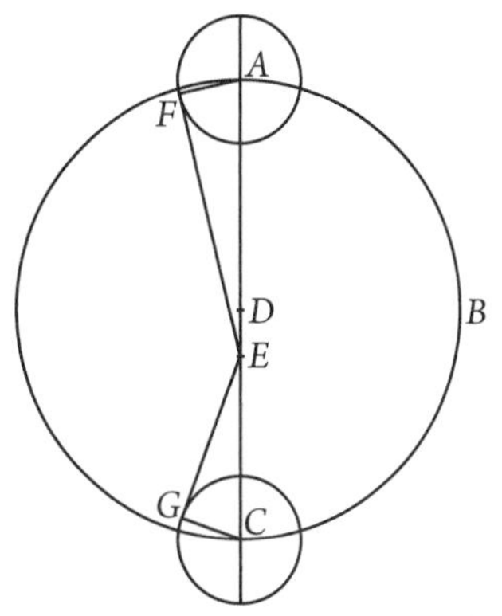

Therefore, since angle *AEF*, at the center of the zodiac circle, subtends the greatest elongation of the star at apogee, assumed to be $44\frac{4}{5}$ degrees, it would

[1] *the eccentric circle carrying the epicycle of the star Venus is stationary*: Venus' eccentric deferent may be called "stationary" only in comparison to Mercury, whose complicated hypothesis required a rapidly moving line of apsides. For Venus and the remaining three planets Ptolemy continues to posit the slow apsidal drift of about 1 degree in 100 years that he first identified in Book IX, Chapter 5.

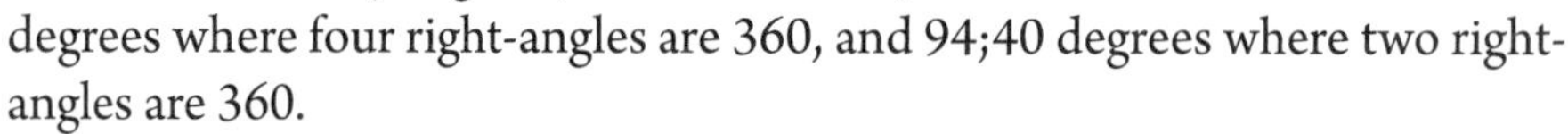

be 44;48 degrees where four right-angles are 360, and 89;36 degrees where two right-angles are 360.

So that also, the arc on chord *AF* is 89;36 degrees where the circle around right-angled triangle *AEF* is 360,[1]
while the chord *AF* subtending it is 84;33 parts, most nearly, where hypotenuse *AE* is 120.

H302

Similarly, since angle *CEG* subtends the greatest elongation at perigee, assumed itself also to be $47\frac{1}{3}$ degrees, it would be 47;20 degrees where four right-angles are 360, and 94;40 degrees where two right-angles are 360.

So that the arc on *CG* is 94;40 degrees where the circle around right-angled triangle *CEG* is 360, while the chord *CG* subtending it is 88;13 parts, most nearly, where hypotenuse *EC* is 120.

And therefore *EC* will be 115;1,
the whole *AC* clearly 235;1,
AD, half of it, 117;30, most nearly,
and the remainder *DE* between the centers 2;29 parts where radius *CG* (that is to say *AF*) of the epicycle is 84;33 and straight line *AE* 120.

So that straight line *DE* between the centers also will be $1\frac{1}{4}$, most nearly, and radius *AF* of the epicycle $43\frac{1}{6}$ parts where radius *AD* of the eccentric circle is 60.

3. On the Ratios of the Eccentricity of the Star.

Observations 10.7–10.8
Calculations 10.7–10.8

H303

Since it is unclear whether the uniform motion of the epicycle is produced around point *D*,[2] we took in this case two greatest elongations in opposite directions with respect to the mean passage of the sun when it ⟨i.e., the planet⟩ is a quadrant removed in either direction from apogee.[3] Of these we observed one in the 18th year of Hadrian, by Egyptian reckoning Pharmouthi 2/3, in which the morning star Venus was at greatest elongation

---

[1] Ptolemy here converts degrees to "demi-degrees," that is, degrees such that 360 make up only two right angles. This technique was discussed earlier in the footnote on page 107 in Book III, chapter 4.

[2] *it is unclear whether the uniform motion of the epicycle is produced around point D*: Point *D* in the previous diagram is the center of circle on which the epicycle is carried. Thus Ptolemy is investigating whether the center of this circle is also the center of regular motion of the epicycle's center.

[3] Ptolemy measures GWE and GEE on two occasions when the mean sun is 90 degrees from apogee—that is, as he is about to state, at $25\frac{1}{2}$ degrees in Aquarius.

from the sun and, being sighted relative to the star called Antares, occupied $11\frac{11}{12}$ degrees of Capricorn, the mean sun then occupying $25\frac{1}{2}$ degrees of Aquarius. So that its greatest morning elongation from the sun's mean passage was $43\frac{7}{12}$ degrees. We observed the other in the 3rd year of Antoninus, by Egyptian reckoning Pharmouthi 4/5, in the evening, in which the star Venus was at greatest elongation from the sun and, being sighted relative to the bright Hyad,[1] occupied $13\frac{5}{6}$ degrees of Aries, the mean sun, again, occupying $25\frac{1}{2}$ degrees of Aquarius. So that in this case, its greatest evening elongation from the sun's mean passage was $48\frac{1}{3}$ degrees.

When these values are assumed, let *ABC* be the diameter through the apogee and perigee of eccentricity, and let *A* be assumed as the point near the 25th degree of Taurus,[2] and *B* the center of the zodiac circle. Let it be proposed to find the center around which we say the uniform motion of the epicycle is produced. Now, let this center be point *D*,[3] and let *DE* be drawn through it at right angles to *AC*, in order that the mean passage of the epicycle be a quadrant distant from apogee,[4] just as in the observations. And let *E* the center of the epicycle (according to the above observations) be taken on it, and when epicycle *FG* is described around it, let tangents to it, *BF* and *BG*, be drawn from *B*, and let *BE*, *EF*, and *EG* be joined.

Proposition 10.3

H304

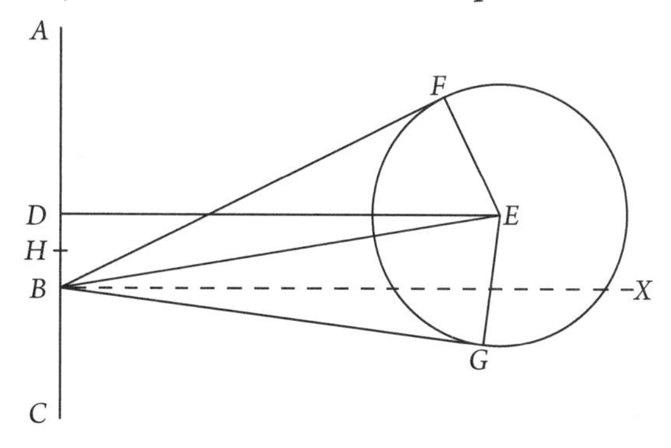

Therefore, since, according to the mean passage in question, the greatest morning elongation from the mean passage is assumed to be $43\frac{7}{12}$ degrees, while the greatest evening elongation $48\frac{1}{3}$ degrees,
the whole angle *FBG* would be 91;55 degrees where four right-angles are 360;

---

[1] *the bright Hyad*: Aldebaran.

[2] *near the 25th degree of Taurus*: that is, apogee, as established in Chapter 1.

[3] *Now, let this center be point D*: The proposed center *D* of regular motion is assumed to be different from *H*, the center of the eccentric deferent.

[4] *in order that the mean passage of the epicycle be a quadrant distance from apogee*: Since the mean longitudinal motions of the sun and the planet are equal, angle *ABX* (between apogee and the mean sun) and angle *ADE* (between apogee and the epicycle's center) will always be equal. Here Ptolemy takes them both as right angles, in accordance with the observations made when the mean sun was a quadrant's distance from apogee. Note: the dotted line *BX*, in the direction of the mean sun, has been added to Ptolemy's original diagram.

and therefore half of it, angle *FBE*, is the same 91;55 degrees where two right-angles are 360.

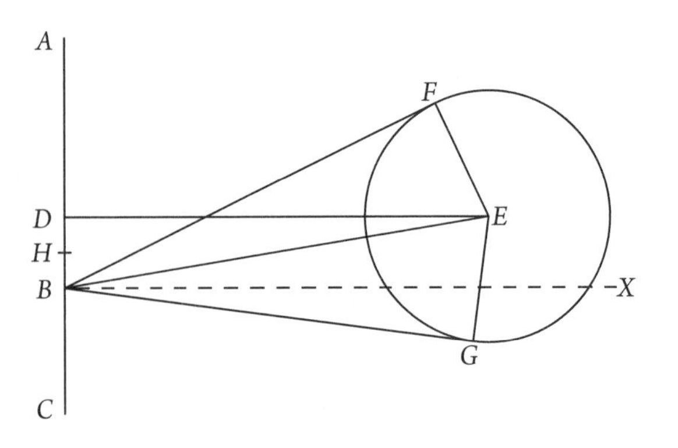

So that also, the arc on *EF* is 91;55 degrees where the circle around right-angled triangle *BEF* is 360, while straight line *EF* itself is 86;16 parts where hypotenuse *BE* is 120.

H305

And therefore *BE* will be 60;3 parts where radius *EF* of the epicycle is 43;10.

Again, since the difference between the above greatest elongations, being 4;45 degrees, contains twice the zodiacal anomalistic difference at that time[1] (which is contained by angle *BED*),
angle *BED* would be 2;22$\frac{1}{2}$ degrees where four right-angles are 360, and 4;45 degrees where two right-angles are 360.

So that also, the arc on *BD* is 4;45 degrees where the circle around right-angled triangle *BDE* is 360,
and straight line *BD* itself is 4;59 parts, most nearly, where hypotenuse *BE* is 120.

And therefore *BD* also will be 2$\frac{1}{2}$ parts, most nearly, where straight line *BE* is 60;3 and the radius of the epicycle 43;10.

The straight line between *B* (the center of the zodiac circle) and the center of the eccentric circle (upon which the center of the epicycle always is) was shown to be 1$\frac{1}{4}$ of the same parts {H302.16};
so that it is half of *BD*.

Therefore, if we bisect *BD* at *H*, we will have it as proven that each of the straight lines *BH* and *HD* between the centers is 1$\frac{1}{4}$ and radius *EF* of the epicycle 43;10 parts where radius *HA* of the eccentric circle carrying the epicycle is 60;
which it was proposed to prove.

H306

---

[1] *twice the anomalistic difference*: If the center of the epicycle moved uniformly, it would lie on the line *BX*, perpendicular to *AC*; and the two angles of elongation, $\angle EBG$ and $\angle EBF$, would be equal. But the center of the epicycle is actually at *E*, and $\angle EBX$ ($= \angle BED$) is the anomalistic difference. The elongation is $\angle XBG$.

Now $\angle XBG = \angle EBG - \angle EBX$; and $\angle XBF = \angle EBF + \angle \mathrm{EBX}$ or $\angle \mathrm{EBG} + \angle \mathrm{EBX}$. Therefore

$$\angle \mathrm{XBF} - \angle \mathrm{XBG} = 2\angle \mathrm{EBX} \text{ or } 2\angle \mathrm{BED};$$

that is, the difference of elongations is twice the anomalistic difference.

4. On the Correction of the Periodic Motions of the Star.

The mode, then, of the hypothesis and the ratios of the anomalies have been taken by us in this way; for the sake of the periodic motions of the star and its epochs we again took two undisputed observations from those of our time and from the ancients.

Observation 10.9, Calculation 10.9

We observed, then, in the 2nd year of Antoninus, by Egyptian reckoning Tubi 29/30, through the astrolabe the star Venus after its greatest morning elongation relative to Spica, and it appeared to occupy $6\frac{1}{2}$ degrees of Scorpio. Then it was in the middle of and also in a straight line with the most northern of the stars in the forehead of Scorpio and the apparent center of the moon, but was $1\frac{1}{2}$ degrees westward of the center of the moon, by which amount it was to the east of the northernmost star in the forehead. But the fixed star then occupied, by our principles of reckoning, 6;20 degrees of Scorpio and is more northerly than the mid-zodiac circle by 1;20 degrees.

H307

The time was $4\frac{3}{4}$ equinoctial hours after mid-night, since, the sun being around 23 degrees of Sagittarius, the 2nd degree of Virgo was culminating in the astrolabe; during this time the sun in its mean passage occupied 22;9 degrees of Sagittarius, while the moon occupied 11;24 degrees of Scorpio, 87;30 degrees of anomaly from apogee, and 12;22 degrees of latitude from the northern limit. And, for these reasons, the center of it occupied precisely 5;45 degrees of Scorpio, but was more northerly than the mid-zodiac circle by 5 degrees, and in Alexandria appeared to occupy 6;45 degrees of Scorpio in longitude, but was more northerly than the mid-zodiac circle by 4;40 degrees. Therefore for these reasons as well, the star Venus occupied 6;30 degrees of Scorpio and was more northerly than the mid-zodiac circle by 2;40 degrees.

Proposition 10.4

When these values are assumed, let *ABCDE* be the diameter through the apogee, and let *A* be assumed at the 25th degree of Taurus, *B* the center around which the epicycle moves uniformly, *C* the center of the eccentric circle (upon which the center of the epicycle is carried) and *D* the center of the zodiac circle.

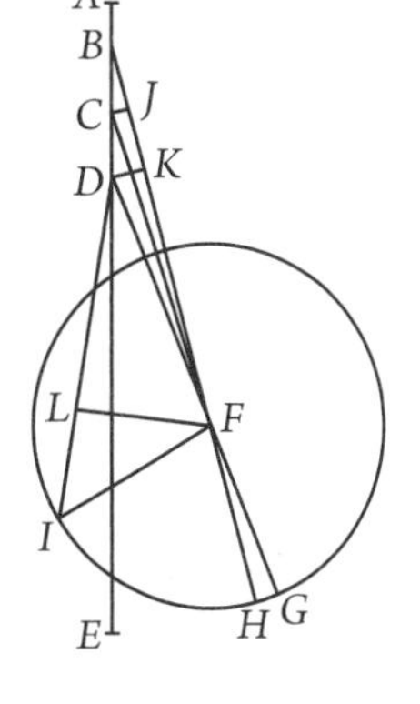

And since the mean sun in the observation occupied 22;9 degrees of Sagittarius, so that the mean passage of the epicycle also was elongated, eastwards, from the perigee at *E* by 27;9 degrees, let its center be assumed at *F*, and when epicycle *GHI* is drawn about it, let *DFG*, *CF*, and *BFH* be joined.

H308

And let *CJ* and *DK* be drawn perpendicular from *C* and *D* to *BF*, and, when the star is assumed at point *I*, let *DI* and *FI* be joined, and let *FL* be drawn perpendicular. Let it be proposed to find the arc *HI*, by which the star was elongated from apogee *H* of the epicycle.

Accordingly, since angle *EBF* is 27;9 degrees of which four right-angles are 360,
and 54;18 of which two right-angles are 360,
the arc on *CJ* also would be 54;18 degrees of which the circle around right-angled triangle *BCJ* is 360,
while the arc on *BJ* is the 125;42 degrees supplementary to the semicircle.

And therefore as to the chords that subtend them:
CJ will be 54;46 parts of which hypotenuse *BC* is 120,
and *BJ* 106;47 of the same.
So that also, *CJ* also will be 0;34
and *BJ*, similarly, 1;7 parts of which straight line *BC* is 1;15 and radius *CF* of the eccentric circle is 60 {H305}.

H309

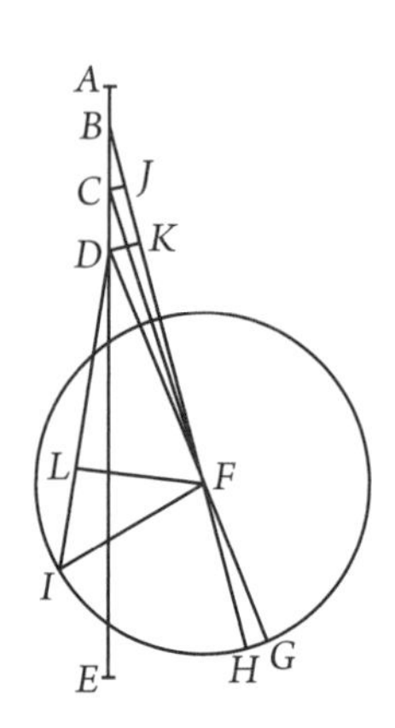

And since the square on *FC* minus the square on *CJ* makes the square on *FJ*,
FJ itself also will be 60, most nearly, of the same.

KJ is also equal to *JB*,
and *DK* is double *CJ*,
because *BC* is also equal to *CD* {H305}.

So that also, *FK* will be the remaining 58;53 parts,
and *DK* 1;8 of the same.

And, for this reason, hypotenuse *FD* also will be 58;54, most nearly.
And therefore *DK* also will be 2;18 parts of which straight line *FD* is 120,
and the arc on it 2;12 degrees of which the circle around right-angled triangle *DFK* is 360.

So that also, angle *BFD* is 2;12 degrees of which two right-angles are 360,
and the whole angle *EDF* 56;30 of the same.

And angle *EDI* also is 18;30 degrees of which four right-angles are 360,
because the star is westwards by so many degrees, according to the observation {H307}, of perigee at *E* (that is to say the 25th degree of Scorpio)
and 37 degrees of which two right-angles are 360.

And therefore the whole angle *IDF* is 93;30 degrees of which two right-angles are 360,
while the arc on *FL* is 93;30 degrees of which the circle around right-angled triangle *DFL* is 360.

H310

And therefore chord *FL* subtending it is 87;25 parts of which *FD* is 120,
but 42;54 parts of which it is 58;54; {H306} that is to say, 42;54 parts of which radius *FI* of the epicycle is 43;10.

So that also, *FL* also will be 119;18 parts of which hypotenuse *FI* is 120,
while the arc on it will be 167;38 degrees of which the circle around right-angled triangle *FIL* is 360.

And therefore angle *FID* is 167;38 of the same degrees of which angle *FDI* also is assumed as 93;30,
and the whole angle *IFG* is 261;8.

Angle *BFD* (that is to say, angle *GFH*) also was shown to be 2;12 of the same;
and therefore the remaining angle *HFI* will be 258;56 degrees of which two right-angles are 360,
and 129;28 of which four right-angles are 360.

Therefore, during the time in question, the star Venus was elongated westwards from *H*, the apogee of the epicycle, the present 129;28 degrees, but eastwards, according to the motion consistent with the hypothesis, the 230;32 degrees supplementary to one circle;
which it was necessary to find.

Observations 10.10-11

Among the ancient observations, we took one which Timocharis records as follows; in the 13th year of Philadelphus, by Egyptian reckoning Mesore 17/18, in the 12th hour, the star Venus appeared to have overtaken the star

H311

situated opposite to Vindemiatrix precisely. And this star, in our time, is beyond the star on the tip of the southern wing of Virgo, and it occupied $8\frac{1}{4}$ degrees of Virgo in the 1st year of Antoninus. Accordingly, since the year of the observation is the 476th from Nabonassar, and the 884th year up to the reign of Antoninus, so that $4\frac{1}{12}$ degrees, most nearly, of the motion of the fixed stars and apogees correspond to the intervening 408 years {H34}, it is clear that the star Venus also occupied $4\frac{1}{6}$ degrees of Virgo, while the perigee of its eccentric circle occupied $20\frac{11}{12}$ degrees of Scorpio.

In this case also, the star Venus passed its greatest morning elongation. For, four days after the present observation, on Mesore 21/22, on the basis of what Timocharis says, it occupied, on our principles of reckoning, $8\frac{5}{6}$ degrees of Virgo, while the mean passage of the sun, according to the earlier observation, occupied 17;3 degrees of Libra, while according to the subsequent observation 20;59 degrees of Libra. So that also, the elongation of the earlier observation is deduced as 42;53 degrees, and that of the subsequent observation as 42;9 degrees.

Proposition 10.5

Now, when these values are given, let the like diagram be set out again, but with the epicycle westwards of perigee, because

H312

the mean passage of the epicycle occupies 17;3 degrees of Libra, and the perigee 20;55 degrees of Scorpio.

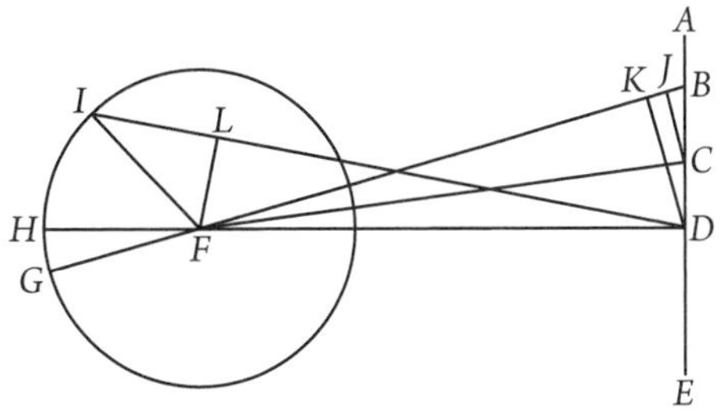

Accordingly, since, for this reason, angle *EBF* is 33;52 degrees of which four right-angles are 360,

and 67;44 of which two right-angles are 360,
the arc on *CJ* also would be 67;44 degrees of which the circle around right-angled triangle *BCJ* is 360,
while the arc on *BJ* would be the 112;16 degrees supplementary to the semicircle.

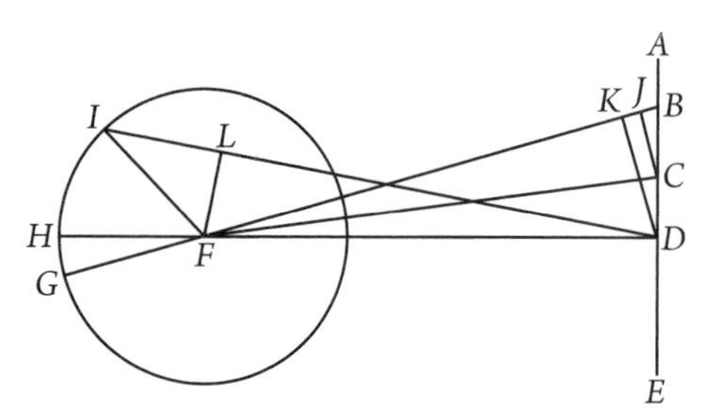

And therefore as to the chords that subtend them:
CJ is 66;52 parts of which hypotenuse *BC* is 120,
and *BJ* 99;38 of the same.

So that also, *CJ* also will be 0;42
and *BJ*, similarly, 1;2 parts of which straight line *BC* is 1;15 and radius *CF* of the eccentric circle 60.

And, since the square on *FC* minus the square on *CJ* makes the square on *FJ*,
FJ itself in length also will be 60, most nearly, of the same.

And for the same reasons {H309ff.}, *BJ* also is equal to *JK*, H313
and *DK* is double *CJ*;
so that also, the remainder *FK* will be 58;58
and *DK* 1;24 of the same.

And, for these reasons, hypotenuse *FD* also is 58;59, most nearly.

And therefore *DK* also will be 2;51 parts of which *FD* is 120,
while the arc on it will be 2;44 degrees of which the circle around right-angled triangle *FDK* is 360.

So that also, angle *BFD* is 2;44 degrees of which two right-angles are 360;
and the whole angle *EDF* will be 70;28 of the same.

Angle *EDI*, by which the star is elongated westwards of perigee, also is 76;45 degrees {H311} of which four right-angles are 360,
and 153;30 of which two right-angles are 360.

So that also, the remaining angle *FDI* is 83;2 of the same,
while the arc on *FL* is 83;2 degrees of which the circle around right-angled triangle *DFL* is 360.

And therefore the chord *FL* subtending it will be 79;33 parts of which hypotenuse *DF* is 120,
and 39;7 of which *DF* is 58;59,
that is to say {H306} 39;7 of which radius *FI* of the eccentric circle is 43;10.

So that also, straight line *FL* will be 108;45 parts of which hypotenuse *FI* is 120,
while the arc on it will be 130 parts, most nearly, of which the circle around right-angled triangle *FIL* is 360.

H314 And therefore angle *DIF* is 130 degrees of which angle *FDI* is assumed to be 83;2,
and the whole angle *HFI* is 213;2 of the same.

Angle *BFD*, that is to say, angle *GFH*, also was shown to be 2;44 of the same;
and therefore the whole angle *GFI* is 215;46 degrees of which two right-angles are 360,
and 107;53 of which four right-angles are 360.

And therefore at this time the star Venus was elongated from apogee *G* of the epicycle, eastwards, by the 252;7 degrees supplementary to one circle; which it was necessary to prove.

Calculation 10.12

It was, accordingly, at the time of our observation similarly elongated from the apogee of the epicycle by 230;32 degrees, and the time-span between the two observations contains 409 Egyptian years and 167 days, most nearly, and 255 whole periodic returns of anomaly; since, as 8 Egyptian years make, most nearly, 5 cycles {H215}, 408 years yield 255 cycles, while the remaining one year, together with the added days, does not complete the time-span of one periodic return. Hence, it has become clear to us that in 409 Egyptian years and 167 days the star Venus, beyond 255 whole periodic returns of anomaly, takes in addition 338;25 degrees H315 on the epicycle; by as many degrees as our epoch exceeded the previous epoch. And this number of degrees, roughly speaking, of surplus is also deduced in the tables of the mean motions previously set out by us {IX.4}, because their correction also arises out of the surplus of cycles that has been found, when the time is reduced to days, and the periodic returns, along with the surplus, into degrees. For when the number of degrees is divided by the number of days, the daily mean movement of anomaly set out by us for the star Venus arises {H216}.

5. On the Epoch of its Periodic Motions.

Calculation 10.13

Since what remains here is to establish the epochs of its periodic motions for the 1st year of the reign of Nabonassar, by Egyptian reckoning Thoth 1, mid-day, we took, again, the time between this year and the year according to the more ancient observation; and this sums up to 475 Egyptian years and $346\frac{3}{4}$ days, most nearly.

And there lies alongside this time-span, along the columns of the anomaly {H238 ff.}, a surplus of mean motion of 181 degrees, most nearly; H316 if we subtract them from the 252;7 degrees according to the observation, we will have as epoch for the 1st year of Nabonassar, by Egyptian reckoning Thoth 1, mid-day: 71;7 degrees of anomaly from apogee of the epicycle; the mean passage of its longitude is assumed equal, again, to that of the sun, that is to say, occupying 0;45 degrees of Pisces {Book III, H257.10}.

It is clear that, since the apogee, according to observation, also occurs around 20;55 degrees of Taurus and $4\frac{3}{4}$ degrees correspond to the intervening 476 years, most nearly {H34}, the apogee will be around 16;10 degrees of Taurus at the time set out for the epoch.

6. Preliminaries to the Demonstrations Concerning the Remaining Stars.

Now, for these two stars, Mercury and Venus, we actually have employed the preceding methods for the conceptions of the hypotheses and the demonstrations of the anomalies. But for the remaining three stars, Mars, Jupiter, and Saturn, we find that the hypothesis of motion is one and the same as that established for the star Venus; that is to say, the hypothesis according to which the eccentric circle (upon which the center of the epicycle is always carried) is described with center bisecting the straight line between the center of the zodiac circle and the center that produces H317 the uniform revolution of the epicycle. This is because, roughly speaking, for each of these three stars the straight line, which is discovered through the greatest difference of zodiacal anomaly,[1] is found to be double, most nearly, the value for the eccentricity established from the size of the retrogradations[2] at the greatest and least elongation of the epicycle.[3] And we find that the demonstrations, through which we establish the sizes of each of the anomalies and the apogees, can no longer be pursued for these three stars in the same way as for those two stars; because they make every elongation from the sun, and because it is not clear from observations (as

---

[1] *the straight line, which is discovered through the greatest difference of zodiacal anomaly*: Ptolemy is referring to the distance between the center of the mid-zodiac circle and the equant. He will assume that for Mars, Jupiter, and Saturn this distance is double the eccentricity of the deferent—just as he found it to be for Venus. His reference to "greatest difference of zodiacal anomaly" finesses the complications of the problem; see the Preliminaries to Book X.

[2] *the eccentricity established from the size of the retrogradations*: This is a puzzling statement, since the eccentricity is determined in Chapter 7, using a procedure that completely eliminates the effect of the epicycle, which is what produces the retrogradations. In fact, the retrogradations are not brought up until Book XII, where Ptolemy shows how to find the magnitude of the retrogradation from the eccentricity and the size of the epicycle. He has no procedure for doing the inverse; that is, for finding the eccentricity from known retrogradations. Perhaps he means by this statement that the correctness of the theory of retrogradations in Book XII can serve as a confirmation of the correctness of the eccentricity found in X.7, determined using successive approximations.

[3] *the greatest and least elongation of the epicycle*: that is, apogee and perigee of the eccentric deferent, where the zodiacal anomaly is zero.

it is for the greatest elongations of the star Mercury and the star Venus) when the star meets the straight line, extended from our sight, tangent to the epicycle. Now, since this approach is not viable, we have employed their diametrically opposite positions as observed relative to the mean passage of the sun. We show from these, first, the ratios of eccentricity and the apogees, since only in the passages observed in this way do we find the zodiacal anomaly distinguished by itself, since then no difference arises due to the heliacal anomaly.

1318

Proposition 10.6

For let eccentric circle ABC of the star, upon which the center of the epicycle is carried, be around center D, and AC its diameter through apogee, and on it point E the center of the zodiac circle, and F of the eccentric circle, relative to which the mean longitudinal passage of the epicycle is observed. And, when epicycle $GHIJ$ is described around B, let $FJBH$ and $GBIEK$ be joined.[1]

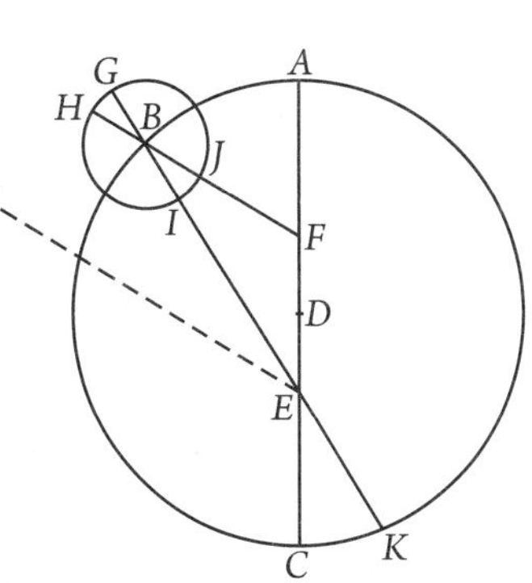

I say, first, that, when the star appears on EG through center B of the epicycle, the mean passage of the sun will always be on the same straight line;

and that when the star is at G,[2] it is in conjunction with the mean passage of the sun (itself also observed at G),

and when it is at I,[3] it will be diametrically opposite to the mean passage of the sun, observed at point K.

For the mean elongations, in longitude and anomaly, from apogees, for each of these three stars, when added together produce the mean passage of the sun from the same starting point.[4]

---

[1] Additionally, draw line EX parallel to FB; its usefulness will become apparent later.

[2] Point G is sighted from the center of the zodiac circle through the center of the epicycle and is therefore "true" apogee of the epicycle. Except at apogee and perigee of the eccentric deferent, it is not in line with the "mean apogee" (point H in this figure) from which the planet's regular motion on the epicycle is measured, as Ptolemy established in Book IX, Chapter 5.

[3] Point I is the true perigee of the epicycle.

[4] *For the mean elongations, in longitude and anomaly, from apogees, ... when added together produce the mean passage of the sun from the same starting point*: This is the relation $s=l+a$ discussed in the Preliminaries to Book IX, where s is the motion of the mean sun, l the planet's mean motion in longitude, and a its mean motion in anomaly. By specifying that the "starting point" for measurement of l, a, and s is the apogee, A, Ptolemy is tacitly assuming that at some time the center of the planet's epicycle, the planet itself, and the mean sun either once were or will be all lined up at A on the planet's line of apsides.

And the angle at B, containing the uniform passage on its epicycle, is always the difference between the angle at center F (which contains the uniform, longitudinal motion of the star) and the angle at E (which contains its apparent motion[1]) [Euc. 1.32].

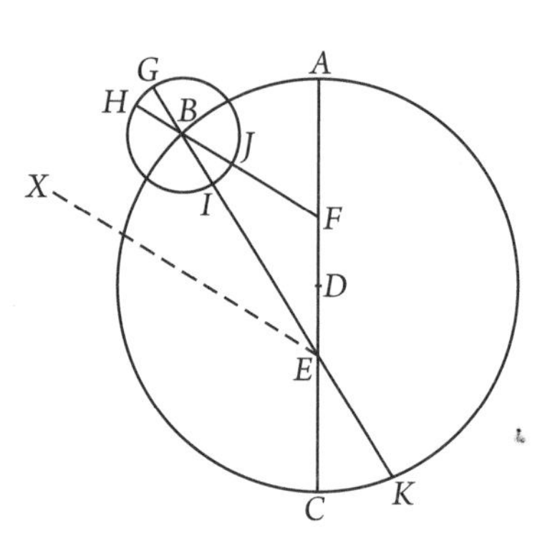

H319

Hence, it is clear that, when the star is at point G, it will fall short of its periodic return to ⟨mean⟩ apogee[2] H by angle GBH, which [Euc. 1.15], when combined with angle AFB, that is to say, subtracted from it, produces angle AEG (contained by the mean solar passage), which is the same as the apparent passage of the star.[3]

When it is at point I, it will, again, have moved on the epicycle angle HBI;

this, when added to angle AFB, will produce the mean passage of the sun from apogee A, containing a semicircle plus angle AFB minus angle JBI: that is to say, angle CEK [Euc. 1.32, Euc. 1.15],

again diametrically opposite to the apparent passage of the star.

Proposition 10.7

And, for this reason, for these configurations,[4] the straight line from center B of the epicycle, when produced to the star, and the line produced from E (our visual center) to the mean passage of the sun, both meet on

---

[1] *its apparent motion*: that is, the apparent motion of the center of the epicycle.

[2] *mean apogee*: The point (here H) from which the planet's regular motion on the epicycle is measured; it is sighted from the equant (here F) through the center of the epicycle, as Ptolemy established in Book IX, Chapter 5. In contrast, point G, sighted from the center of the zodiac circle through the epicycle's center, is the *true apogee* of the epicycle. At apogee and perigee of the eccentric deferent, the mean and true apogees coincide.

[3] This paragraph may be paraphrased: Suppose the planet and the epicycle begin from their respective apogees (H and A). As the center B of the epicycle moves (counterclockwise) through the angle AFB around the equant, the planet moves (also counterclockwise) around the epicycle from H through I and J to G, where it is again in conjunction with center B of the epicycle. By the relationship $s = l + a$ noted above, the mean sun's motion about E must be the sum of these angles, that is, angle AFB plus an angle that is almost a complete circle, falling short by angle GBH. Thus the mean sun, starting from A, moves counterclockwise around E through an angle AEX equal to angle AFB, and then through the reflex angle HBG, ending at B. The effect is *as if* the acute angle HBG, or GEX equal to it, were subtracted from angle AFB, or AEX equal to it; hence the phrase, "that is to say, subtracted from it."

[4] *for these configurations*: that is, when the star is at either G or I on the epicycle.

H320 one and the same straight line; but for all the remaining elongations, they make inclinations that are different, but always parallel to one another.[1]

For if, at whatever position in the present diagram, we draw from *B* to the star a straight line, as *BL*, and from *E* to the mean passage of the sun, as *EM*,
angle *AEM*, because of what was previously stated {H319ff.}, will be equal to angle *AFH* plus angle *LBH*,
while angle *AFH* also will be equal to angle *AEG* plus angle *GBH* [Euc. 1.32, Euc. 1.15].

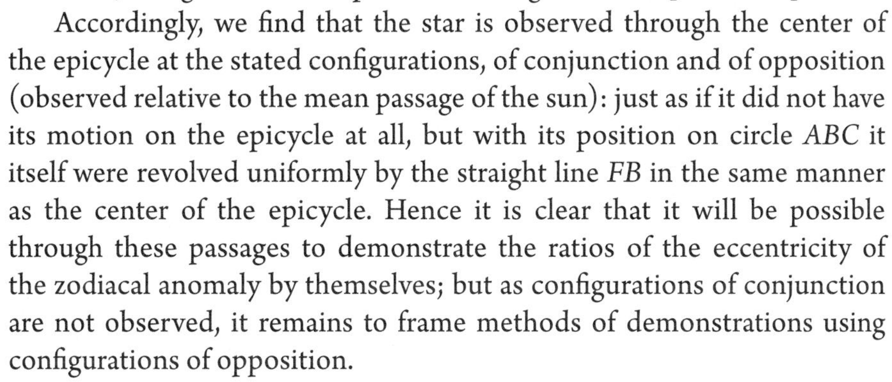

When angle *AEG* is subtracted in common, then the remaining angle *GEM* will be equal to the remaining angle *GBL*;
therefore, straight line *EM* is parallel to straight line *BL* [Euc. 1.28].

Accordingly, we find that the star is observed through the center of H321 the epicycle at the stated configurations, of conjunction and of opposition (observed relative to the mean passage of the sun): just as if it did not have its motion on the epicycle at all, but with its position on circle *ABC* it itself were revolved uniformly by the straight line *FB* in the same manner as the center of the epicycle. Hence it is clear that it will be possible through these passages to demonstrate the ratios of the eccentricity of the zodiacal anomaly by themselves; but as configurations of conjunction are not observed, it remains to frame methods of demonstrations using configurations of opposition.

7. Demonstration of the Eccentricity and the Apogee of the Star Mars.

By taking the places and times of three full-moon eclipses we geometrically proved both the ratio of anomaly and the place of apogee for the moon {IV.6}. In just the same way, then, here also we observed as precisely as possible (through astrolabe instruments) the places of three oppositions, diametrically opposite to the mean passage of the sun, for each of these stars. And we calculated in addition (from the mean passages of the sun during H322 the observations) the time and place of the separation rather minutely. From these we prove both the ratio of the eccentricity and the apogee.

---

[1] *always parallel to one another*: For example, *BL* will always be parallel to *EM*, as Ptolemy will outline in the following paragraph.

Observations 10.12–14
Calculation 10.14

Accordingly, for the star Mars first, we took three oppositional configurations.[1] We observed the first of these in the 15th year of Hadrian, by Egyptian reckoning Tubi 26/27, one equinoctial hour after mid-night, around 21 degrees of Gemini, the second in the 19th year of Hadrian, by Egyptian reckoning Pharmouthi 6/7, three hours before midnight, around 28;50 degrees of Leo, and the third in the 2nd year of Antoninus, by Egyptian reckoning Epiphi 12/13, two equinoctial hours before midnight, around 2;34 degrees of Sagittarius. The time-spans, then, of the intervals contain, from the first opposition to the second, 4 Egyptian years, 69 days, and 20 equinoctial hours, and from the second to the third, similarly, 4 years, 96 days, and 1 equinoctial hour. From the time-span of the first interval, beyond whole cycles of motion of longitude, 81;44 degrees are deduced {H234ff.}, and from the second time-span 95;28 degrees. For there will be no significant difference even if we calculate the mean motions from periodic returns set out in a rather rough way, in light of such a time-span. It is clear that the apparent star also has moved, during the first interval, beyond whole cycles, 67;50 degrees, and during the second 93;44 degrees.[2] H323

Proposition 10.8 Part A

H323.10

Now, let there be described in the plane of the zodiac circle three equal circles; of these let the one carrying the center of the epicycle of the star Mars be *ABC* around center *D*, *EFG* the eccentric circle of uniform motion around center *H*, and *IJK* the circle concentric with the zodiac circle around center *L*. And let diameter *MNOP* be through all the centers.

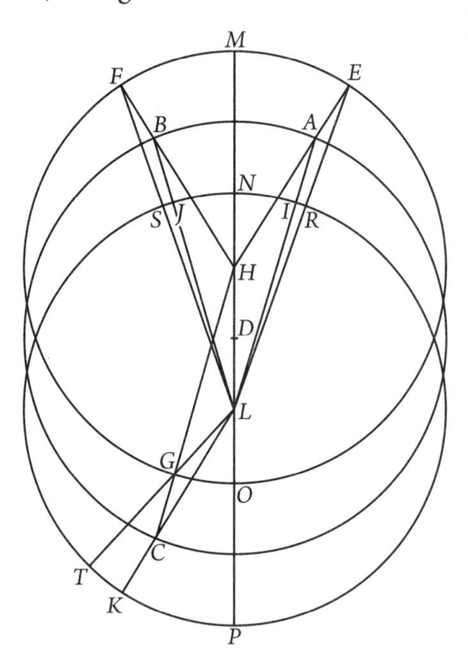

And let *A* be assumed as the point on which the center of the epicycle was at the first opposition, and *B* as the point on which it was at the second opposition, and *C* as the point on which it was at the third opposition.

---

[1] *three oppositional configurations*: In Propositions 10.8–10.13 to follow, Ptolemy will make repeated reference to these three oppositions, both collectively and sequentially.

[2] *the apparent star also has moved, during the first interval, ... 67;50 degrees*: that is, the distance from the star's first opposition at 21° Gemini to its second at 28° 50′ Leo. Similarly, 93;44 degrees is the distance from 28° 50′ Leo to the third opposition at 2° 34′ Sagittarius.

And let *HAE, HBF, HGC, LIA, LJB,* and *LCK* be joined, so that arc *EF* of the eccentric circle is the 81;44 degrees of the first periodic interval, *FG* the 95;28 degrees of the second, and, again, arc *IJ* of the zodiac circle the 67;50 degrees of the first apparent interval, and *JK* the 93;44 degrees of the second. [1]

H324

If, then, arcs *EF* and *FG* of the eccentric circle were subtended by arcs *IJ* and *JK* of the zodiac circle, we would no longer be seeking for anything else to demonstrate the eccentricity.

But they themselves subtend arcs *AB* and *BC* of the eccentric circle in the middle that are not given.

And if we join *LRE, LSF,* and *LGT*, then arcs *RS* and *ST* of the zodiac circle (these themselves clearly not being given) again subtend arcs *EF* and *FG* of the eccentric circle.

Hence, there will have to have been given previously the different segments *IR, JS,* and *KT*, in order that, from the arcs that are in syzygy, *EFG* and *RST*, the ratio of the eccentricity be precisely demonstrated.

It is not possible precisely to take these arcs at all prior to the ratio of eccentricity and the apogee; they will, however, be given most nearly, even if those are not precisely found beforehand, because their differences are not great. Hence, we will first produce the calculation as if the arcs *IJ, JK* differed in nothing significant compared to arcs *RS, ST*.[2]

[1] It is somewhat misleading of Ptolemy to differentiate at this stage the arc *IJ* from arc *RS*, and arc *JK* from arc *ST*. In the following proposition 10.8, he in effect supposes, as a first approximation, that points *A* and *B* coincide with points *E* and *F*, respectively, allowing him to determine the position of point *H*. Then, supposing that *D* lies halfway between *H* and *L*, he determines the corrections *IR, JS,* and *KT*, that are required to locate points *A, B,* and *C* on the eccentric circle with center *D*. This provides new values for the arcs *RS* and *ST*, with which the entire computation must be repeated. For a summary of the full procedure, see the Preliminaries to Book X, above.

[2] *as if the arcs IJ, JK, differed in nothing significant...*: Following Toomer (p. 486 n. 37), this wording corrects an error in Heiberg's text. The consequence of this supposition is that the apparent magnitudes of arcs *AB* and *BC* are the same as those of the arcs *EF* and *FG*, respectively, which requires *A* to coincide with *E, B* with *F*, and *G* with *C*. Ptolemy explicitly acknowledges this in the next sentence.

Proposition 10.8 Part B

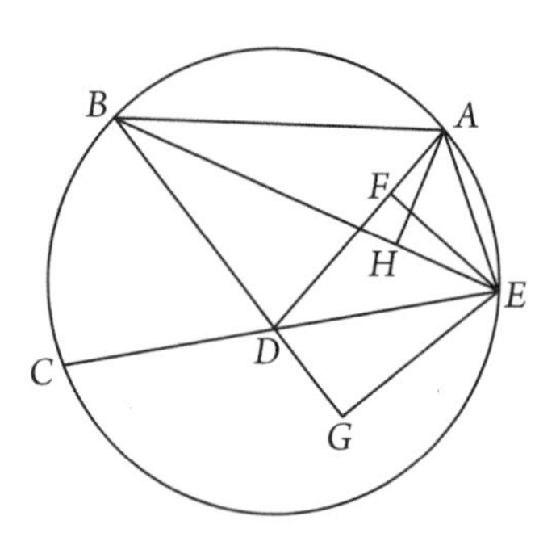

H325

For let ABC be the eccentric circle of the uniform passage of the star Mars,[1] and let A be assumed as the point of the first opposition, B that of the second, and C that of the third. And, within it, let D be taken as the center of the zodiac circle (upon which is our sight),[2] and let straight lines be joined, always from the three points of the oppositions to the point of sight, as in the present case AD, BD, and CD. And, in general, let one of the three joined straight lines be produced to the opposite arc of the eccentric circle, as, in this case, CDE, and let a straight line join the two remaining points of the oppositions, as AB for these. Then, from the section of the eccentric circle produced by the extended straight line, for example E, let straight lines be joined to the remaining two points of the oppositions, as here, EA and EB. And let perpendiculars be drawn ⟨from the endpoint of the extended straight line⟩ to the straight lines joined from the two stated points to the center of the zodiac circle, as, for these, EF to AD, and EG to BD. And, furthermore, from one of the two stated points let there be drawn a perpendicular to the straight line joined from the other of them to the extra point of the eccentric that was produced; as, in this case, straight line AH from A to BE.

H326

Always observing, then, these considerations in this sort of diagram, in whatever manner we like, we will find that the same ratios are maintained numerically,[3] while the remaining proof from the arcs in question for the star Mars will be clear in the following way.

Proposition 10.8 Part C

For since arc BC of the eccentric circle is assumed as subtending 93;44 degrees of the zodiac circle,

angle BDC (at the center of the zodiac circle) would be 93;44 degrees of which four right-angles are 360, and 187;28 of which two-right angles are 360;

and the angle adjacent to it, EDG, would be 172;32 of the same.

---

[1] *the eccentric circle of the uniform passage of the star Mars*: For the first iteration, Ptolemy supposes the the "circle carrying the center of the epicycle" is the same as the "eccentric circle of uniform motion." Both are represented by circle ABC. Later he will correct for this oversimplification (Prop. 10.14).

[2] *let D be taken as the center of the zodiac circle*: Thus D in the present diagram corresponds to L—not D—of the previous diagram.

[3] *the same ratios are maintained numerically*: That is, it makes no difference which of the lines AD, BD, CD is extended to the circumference.

So that also, the arc on *EG* is 172;32 degrees of which the circle around right-angled triangle *DEG* is 360,
while chord *EG* is 119;45 parts of which hypotenuse *DE* is 120.

Similarly, since arc *BC* is 95;28 degrees, angle *BEC* (on the arc) also would be 95;28 degrees of which two right-angles are 360.

Angle *BDE* also was 172;32 of the same;
H327 and therefore the remaining angle *EBG* will be 92 of the same.

So that also, the arc on *EG* is 92 degrees of which the circle around right-angled triangle *BEG* is 360,
while chord *EG* is 86;19 parts of which hypotenuse *BE* is 120.

And therefore *BE* also will be 166;29 parts of which *EG* was shown to be 119;45 and *ED*, similarly, 120.

Proposition 10.8 Part D

Again, since the whole arc *ABC* of the eccentric circle is assumed as subtending[1] the summed up 161;34 degrees of both separations taken together, angle *ADC* also would be 161;34 degrees of which four right angles are 360, while the remaining angle *ADE* would be 18;26 of the same, but 36;52 degrees of which two right angles are 360.

So that also, the arc on *EF* is 36;52 degrees of which the circle around right-angled triangle *DEF* is 360,
and chord *EF* is 37;57 parts of which hypotenuse *DE* is 120.

Similarly, since arc *ABC* of the eccentric circle sums up to 177;12 degrees, angle *AEC* also would be 177;12 degrees of which two right-angles are 360.

Angle *ADE* also was 36;52 of the same; and therefore the remaining angle *DAE* is 145;56 of the same.

So that also, the arc on *EF* is 145;56 degrees of which the circle around right-angled triangle *AEF* is 360,
H328 while chord *EF* is 114;44 parts of which hypotenuse *AE* is 120.

And therefore *AE* also will be 39;42 parts of which *EF* was shown to be 37;57 and the straight line *ED* 120.

Proposition 10.8 Part E

Again, since arc *AB* of the eccentric circle is 81;44 degrees,
angle *AEB* also would be 81;44 degrees of which two right-angles are 360.

So that also, the arc on *AH* is 81;44 degrees of which the circle around right-angled triangle *AEH* is 360,

[1] *subtending the summed up 161;34 degrees*: Arc *ABC subtends* an angle of 161;34 degrees, but is not *measured by* that angle (because the center of the arc is not *D*). Since arc *ABC* is the equant circle, it is measured by the *times* of the observations. The magnitudes of the component arcs *AB* and *BC* are given in Calculation 10.14 {H323}, where they are confusingly labelled *EF* and *FG*. Their sum is 177;12 degrees. One mus be careful to distinguish this arc *ABC*, on the equant circle, from the arc *ABC* in the diagram to Prop. 10.8 Part A {H323.10}, which is the (at this point unknown) arc on the planet's eccentric circle.

while the arc on *EH* is the 98;16 degrees supplementary to the semicircle.

And therefore as to the chords that subtend them:
AH will be 78;31 parts of which hypotenuse *AE* is 120,
and *EH* 90;45 of the same;
so that also, *AH* will be 25;58
and *EH*, similarly, 30;2 parts of which *AE* was shown to be 39;42 and *DE* is assumed to be 120.

The whole *EB* also was shown to be 166;29 of the same;
and therefore the remainder *BH* is 136;27 parts of which *AH* was 25;58.

And the square on *BH* is 18,615;16, while the square on *AH*, similarly, is 674;16,
which, when added together, make the square on *AB* 19,289;32;
therefore, *AB* in length is 138;53 parts of which *ED* was 120 and straight line *AE* 39;42.

H329

And chord *AB* also is 78;31 parts of which the diameter of the eccentric circle is 120;
for it subtends an arc of 81;44 degrees.

And therefore *ED* also will be 67;50 and AE 22;44 of the same parts of which chord *AB* is 78;31
and the diameter of the eccentric circle is 120.

So that also, the arc on it[1] of the eccentric circle is 21;41 degrees,
and the whole arc *EABC* is 198;53 degrees.

And therefore the remaining arc *CE* is 161;7 degrees,
and *CDE* the chord subtending it is 118;22 parts of which the diameter of the eccentric circle is 120.

Proposition 10.9 Part A

If, then, straight line *CE* were found equal to the diameter of the eccentric circle, it is clear that its center would happen to be on *CE* also, and the ratio of the eccentricity would immediately appear; but since *CE* is not equal, but it has made segment *EABC* larger than a semicircle, it is clear that the center of the eccentric circle will fall in this ⟨segment⟩.

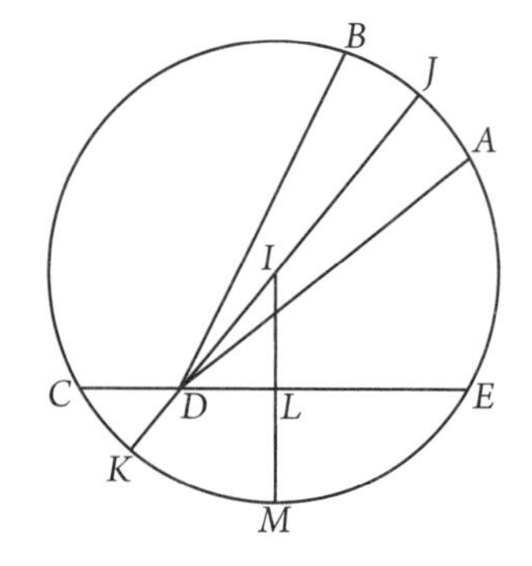

Now, let ⟨the center of circle *ABC*⟩ be assumed to be *I*, and let there be drawn

[1] *the arc on it*: That is, the arc on chord *AE*.

through this and *D* diameter *JIDK* through both centers, and from *I* let *ILM* be drawn perpendicular to *CE*.

H330 Accordingly, since straight line *EC* was shown to be 118;22 parts of which diameter *JK* is 120,
and straight line *DE* also was 67;50 of the same,
therefore the remainder *CD* will also be 50;32 of the same.

So that, since the rectangle contained by *ED,DC* is equal to the rectangle contained by *JD,DK* [Euc. 3.35],
we will have the rectangle contained by *DJ,DK* as 3,427;51 parts.

But the rectangle contained by *JD,DK* plus the square on *DI* also makes the square on half of the whole, that is to say *JI* [Euc. 2.5].

Therefore, if from the square on the half of the resultant 3,600 we subtract the rectangle contained by *JD,DK*, 3,427;51,
there will remain for us the square on *DI*, 172;9 of the same.

And therefore we will have in length the line *DI* between the centers as 13;7 parts,[1] most nearly, of which radius *IJ* of the eccentric circle is 60.

Proposition 10.9 Part B

Again, since half of *CE*, that is to say *CL*, is 59;11 parts of which diameter H331 *JK* is 120,
and straight line *CD* also was shown to be 50;32 of the same,
therefore the remainder *DL* also is 8;39 parts of which *DI* was found to be 13;7.

So that also, *DL* also will be 79;8 parts of which hypotenuse *DI* is 120,
while the arc on it will be 82;30 degrees of which the circle around right-angled triangle *DIL* is 360.

And therefore angle *DIL* is 82;30 degrees of which two right-angles are 360, and 41;15 of which four right-angles are 360.

And since it is at the center of the eccentric circle, we will have arc *KM* also as 41;15 degrees.

And the whole *CKM*, being half of *CME*, also is 80;34;
and therefore the remainder *CK*, from the third opposition to perigee, is 39;19 degrees.

And it is clear that, when *BC* also is assumed as 95;28 degrees, then the remainder *JB*, from apogee to the second opposition, also will be 45;13 degrees;
and, when *AB* is assumed as 81;44 degrees, then the remainder *AJ*, from the first opposition to apogee, also will be 36;31 degrees.

---

[1] *13;7 parts*: This is the first approximation for *DI*, resulting from Ptolemy's first iteration; see "Zodiacal Anomaly and the Two Eccentricities" in the Preliminaries to Book X. Bear in mind that, as footnote 2 on page 180 explained, *DI* here corresponds to what was earlier represented *HL*.

Accordingly, when these values are assumed, let us examine in the following way the differences deduced from them for the arcs being sought of the zodiac circle at each opposition. H332

Proposition 10.10

For, from the above figure of the three oppositions {H323.10}, let the diagram of the first opposition alone[1] be set out, and, when *AD* is joined in addition, let *DU* and *LV* be drawn perpendicular from points *D* and *L* to *AH* produced.[2]

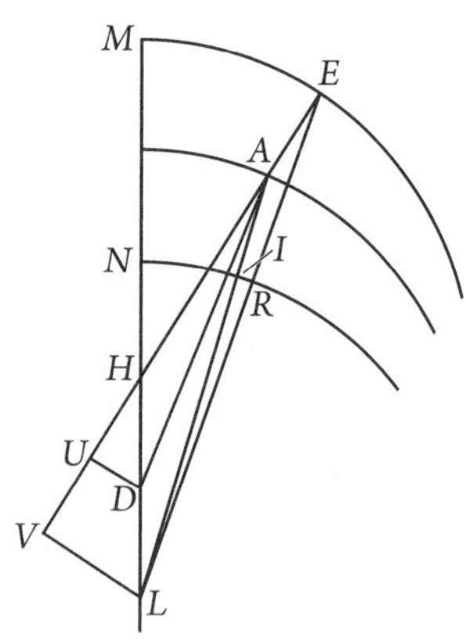

⟨first opposition⟩

Accordingly, since arc *ME* is 36;31 degrees, angle *EHM* also would be 36;31 degrees of which four right-angles are 360,
but it itself and the angle at its vertex, *DHU*, would be 73;2 degrees of which two right-angles are 360;
so that also, the arc on *DU* is 73;2 degrees of which the circle around right-angled triangle *DHU* is 360,
while the arc on *HU* is the 106;58 degrees supplementary to the semicircle.

And therefore as to the chords that subtend them:
DU is 71;25 parts of which hypotenuse *DH* is 120,
while *UH* is 96;27 of the same.

So that also, *DU* also will be 3;54
and *UH*, similarly, 5;16 parts of which straight line *DH* is 6;33$\frac{1}{2}$[3] and radius *DA* of the eccentric circle 60. H333

And since the square on *DU* subtracted from the square on *DA* makes the square on *AU*,
AU also in length will be 59;52,
and the whole *AV* (since *VU* is equal to *UH*) 65;8 parts,
of which *LV*, being double *DU*, is deduced as 7;48.

And thereby hypotenuse *AL* also will be 65;36 of the same.

And therefore *LV* also will be 14;16 parts of which straight line *AL* is 120,

---

[1] *the diagram of the first opposition alone*: The portion of the diagram at H323.10 containing point *A*, as shown here.

[2] The points in the diagrams for Props. 10.10–10.14 have the same labels as those in the diagram for Prop. 10.8; thus *H* in this series corresponds to *I* in Prop. 10.9, and both represent the center of the epicycle. Similarly, *L* in the present series corresponds to *D* in Prop. 10.9. There is no point in Prop. 10.9 corresponding to the present series' point *D*.

[3] *DH is* 6;33$\frac{1}{2}$: Ptolemy here assumes that for Mars, as for Venus, *D* is the midpoint between *H* and *L*. Since *HL* was found to be 13;7 parts (Prop. 10.9 Part A), *DH* will be 6;33$\frac{1}{2}$ parts.

while the arc on it will be 13;40 degrees of which the circle around right-angled triangle *ALV* is 360.

So that also, angle *LAV* is 13;40 degrees of which two right-angles are 360.

Again, since *LV* was also shown to be 7;48

and *VH*, similarly, 10;32 parts of which radius *EH* of the eccentric circle is 60,

then the whole *VHE* will be 70;32 of the same,

and thereby hypotenuse *EL* also will be 71, most nearly.

And therefore straight line *LV* also will be 13;10 parts of which straight line *LE* is 120,

while the arc on it will be 12;36 degrees of which the circle around right-angle triangle *ELV* is 360;

H334 so that also, angle *LEV* is 12;36 degrees of which two right-angles are 360.

Angle *LAV* also was 13;40 of the same;

and therefore the remaining angle *ALE* is 1;4 degrees of which two right-angles are 360, and 0;32 of which four right-angles are 360.

Therefore, arc *IR* of the zodiac circle also is so many degrees.

Proposition 10.11

Now, let the same figure, containing the diagram of the second opposition, be set out {H323.10}.

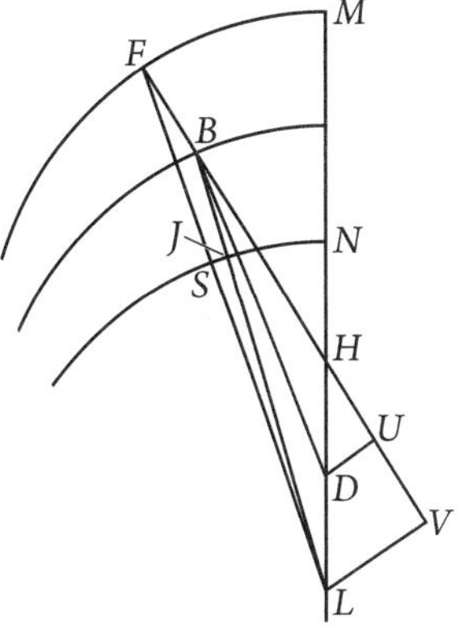

⟨second opposition⟩

Accordingly, since arc *MF* is assumed as 45;13 degrees, angle *MHF* also would be 45;13 degrees of which four right-angles are 360,

but it itself and angle *DHU* at its vertex would be 90;26 degrees of which two right-angles are 360;

so that also, the arc on *DU* is 90;26 degrees of which the circle around right-angled triangle *DHU* is 360,

while the arc on *UH* is the 89;34 degrees supplementary to the semicircle.

And therefore as to the chords that subtend them:

DU is 85;10 parts of which hypotenuse *DH* is 120,

and *UH* is 84;32 of the same.

So that also, *DU* also will be 4;39

H335 and *UH*, similarly, 4;38 parts of which straight line *DH* is $6;33\frac{1}{2}$ and radius *DB* of the eccentric circle 60.

And since the square on *DU*, subtracted from the square on *DB*, makes the square on *BU*,

BU also in length will be 59;49,

and the whole *VB* (because *UV* is equal to *UH*) will be 64;27 parts of which *LV*, being double *DU*, is deduced as 9;18.

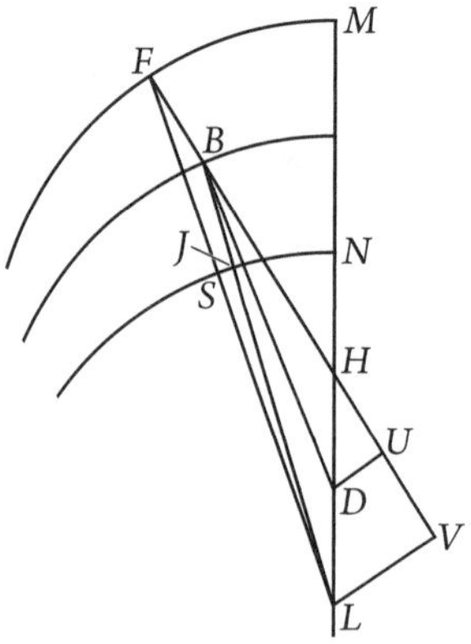

And thereby hypotenuse *LB* also will be 65;6 of the same.[1]

And therefore *LV* also will be 17;9 parts of which *LB* is 120,
while the arc on it will be 16;26 degrees of which the circle around right-angled triangle *BLV* is 360.

So that also, angle *LBV* is 16;26 degrees of which two right-angles are 360.

M
F
B
J
N
S
H
U
D
V
L

⟨second opposition⟩

Again, since *LV* also was shown to be 9;18 and *VH*, similarly, 9;16 parts of which radius *FH* of the eccentric circle is 60, then the whole *VHF* will be 69;16 of the same,
and thereby hypotenuse *LF* also will be 69;52.

And therefore *LV* also will be 16 parts, most nearly, of which hypotenuse *LF* is 120,
while the arc on it will be 15;20 degrees of which the circle around right-angled triangle *FLV* is 360;
so that also, angle *LFV* is 15;20 degrees of which two right-angles are 360. H336

And angle *LBV* also was 16;26 of the same;
and therefore the remaining angle *BLF* is 1;6 of the same and 0;33 of which four right-angles are 360.

Therefore, arc *JS* of the zodiac circle also is so many degrees.

Proposition 10.12

Accordingly, since for the first opposition we have found *IR* as 0;32, it is clear that the first separation, observed relative to the eccentric circle, will be greater than the apparent by the 1;5 parts of both arcs taken together and will contain 68;55 degrees.[2]

Proposition 10.13

Now, let the diagram of the third opposition also be set out {H323.10}.

Accordingly, since arc *OG* also is assumed as 39;19 degrees, angle *OHG* also would be 39;19 degrees of which four right-angles are 360, and 78;38 of which two right-angles are 360.

So that also, the arc on *DU* is 78;38 degrees of which the circle around right-angled triangle *DHU* is 360,
while the arc on *HU* is the 101;22 degrees supplementary to the semicircle.

And therefore as to the chords subtending them:

---

[1] *65;6 of the same*: Following Toomer, this value has been corrected from 69;6.

[2] *68;55 degrees*: Since arc *IJ* was observed to be 67;50° (Obss. 10.12–14 and Calc. 10.14), while arc *IR* has just been found to be 0;32° and arc *JS* to be 0;33°, arc *RS*, the sum of all these, will be 68;55°.

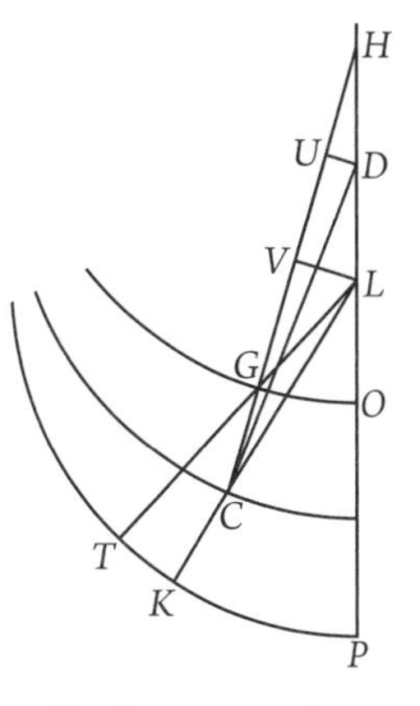

⟨third opposition⟩

DU is 76;2 parts of which hypotenuse DH is 120,
and HU 92;50 of the same;
so that also, DU will be 4;9
and HU, similarly, 5;4 parts of which straight line
H337 DH between the centers is 6;33½ and radius DC of the eccentric circle 60.

And since the square on DU subtracted from the square on CD makes the square on CU,
straight line CU also will be 59;51
and the remainder CV (because HU is equal to UV) will be 54;47 parts
of which LV, being double DU, also is deduced as 8;18.

And thereby hypotenuse LC also is 55;25 of the same.

And therefore LV also will be 17;59 parts of which LC is 120,
while the arc on it will be 17;14 degrees of which the circle around right-angled triangle CLV is 360;
so that also, angle LCV is 17;14 degrees of which two right-angles are 360.

Again, since LV also was shown to be 8;18
and HV, similarly, 10;8 parts of which radius HG of the eccentric circle is 60,
then the remainder VG will be 49;52 of the same,
H338 and thereby hypotenuse LG also will be 50;33.

And therefore LV also will be 19;42 parts of which LG is 120,
while the arc on it will be 18;54 degrees of which the circle around right-angled triangle GLV is 360;
so that also, angle LGV is 18;54 degrees of which two right-angles are 360.

Angle LCV also was shown to be 17;14 of the same;
and therefore the remaining angle CLG is 1;40 of the same, and 0;50 of which four right-angles are 360.

Therefore, arc KT of the zodiac circle also is so many degrees.

Proposition 10.14

Accordingly, since for the second opposition we have found JS as 0;33, it is clear that the second separation, observed relative to the eccentric circle, will be less than the apparent by the 1;23 parts of both arcs taken together and will contain 92;21 degrees.[1]

---

[1] *the second separation ... will be 92;21 degrees*: The apparent separation, 93°44′ {H323}, arc JK diminished by arcs KT (0°50′) and JS (0°33′), leaves 92°21′ as the arc ST.

Summarizing, then, at this point Ptolemy has established the following arcs: EF = 81;44°, FG = 95;28°, IJ = 67;50°, and JK = 93;44° (Obss. 10.12–14 and Calc. 10.14), IR = 0;32° (Prop. 10.10), JS = 0;33° (Prop 10.11), RS = 68;55° (Prop. 10.12), KT = 0;50° (Prop. 10.13) and ST = 92;21° (Prop. 10.14).

Proposition 10.15

For these arcs of the two separations of the zodiac circle that we deduced and for the arcs again naturally assumed on the eccentric circle we followed the theorem proven previously {Props. 10.8–10.9}, through which we demonstrate the apogee and ratio of eccentricity.

Accordingly, we find (in order that we not make our explanation lengthy by repetition) that the straight line *DI* between the centers is 11;50 parts[1] of which the radius of the eccentric circle is 60, H339
while arc *CK* of the eccentric circle, that is to say the arc from the third opposition to perigee, is 45;33 degrees,
on the basis of which, again, arc *JB* also is 38;59
and *AJ*, similarly, 42;45 degrees.

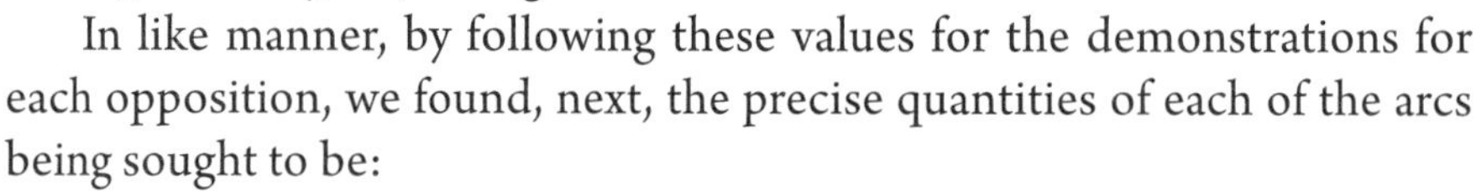

In like manner, by following these values for the demonstrations for each opposition, we found, next, the precise quantities of each of the arcs being sought to be:
IR 0;28,
JS likewise the same 0;28, most nearly,
and *KT* 0;40.

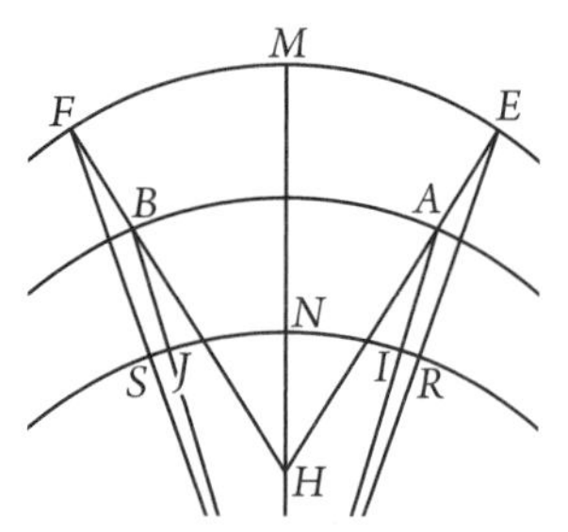

By adding together, of these arcs, the minutes of the first and second opposition and by adding the resultant 0;56 minutes to the 67;50 degrees of the first separation of the zodiac circle,
we obtained the separation precisely observed relative to the eccentric circle as 68;46 degrees.

And by adding together the minutes of the second and third opposition and subtracting the resultant 1;8 degrees from the apparent 93;44 degrees of the zodiac circle for the second separation,
we found the separation, precisely observed, again, relative to the eccentric circle, to be 92;36 degrees.

Next, by using the same demonstration on the basis of these values, we precisely determined both the ratio of eccentricity and the apogee, and H340

[1] *11;50 parts*: This is the second approximation for *DI* (corresponding to *HL* in several other diagrams) resulting from Ptolemy's second iteration.

found that the straight line *DI* between the centers is 12 parts,[1] most nearly,
of which radius *IJ* of the eccentric circle is 60,
and that arc *CK* of the eccentric circle is 44;21 degrees;
on the basis of this arc, again, *JB* also is 40;11 degrees
and *AJ*, similarly, 41;33.

Next, we will show through the same procedures that the observed, apparent separations of the three oppositions are also found in agreement with these quantities.

Proposition 10.16

For let the diagram of the first opposition be set out, containing only eccentric circle *EF*, upon which the center of the epicycle is always carried.

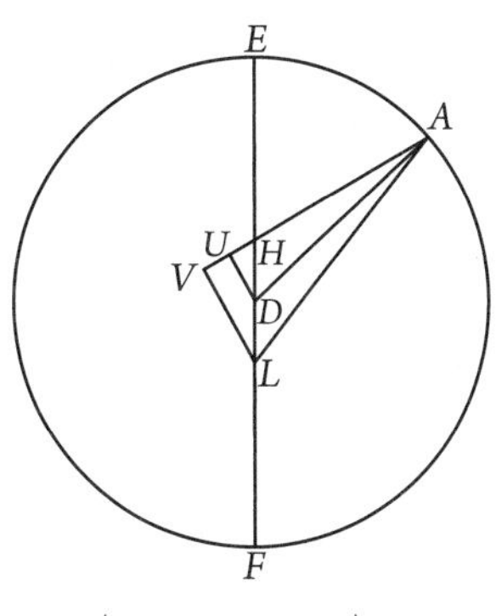

⟨first opposition⟩

Accordingly, since angle *AHE* is 41;33 degrees of which four right-angles are 360,
but both itself and angle *DHU* at its vertex are 83;6 degrees of which two right-angles are 360,
the arc on *DU* also would be 83;6 degrees of which the circle around right-angled triangle *DHU* is 360,
while the arc on *UH* would be the 96;54 degrees supplementary to the semicircle.

And therefore as to the chords that subtend them:
DU is 79;35 parts of which hypotenuse *DH* is 120,
while *UH* is 89;50 of the same;
so that also, *DU* also will be 3;58$\frac{1}{2}$
and *UH*, similarly, 4;30 parts of which straight line *DH* is 6 and hypotenuse *DA* 60.

H341

And, since the square on *DU* subtracted from the square on *DA* makes the square on *UA*,
UA too in length will be 59;50 of the same.

Again, since *UH* is equal to *UV*,
and *LV* is double *DU*,
then we will have the whole *AV* as 64;20 parts of which straight line *LV* is 7;57.

And thereby hypotenuse *LA* also will be 64;52 of the same;
so that also, *LV* also will be 14;44 parts of which straight line *LA* is 120,
while the arc on it will be 14;6 degrees of which the circle around right-angled triangle *ALV* is 360.

[1] *12 parts*: This Ptolemy's third and final approximation for *DI* (corresponding to *HL* in several other diagrams).

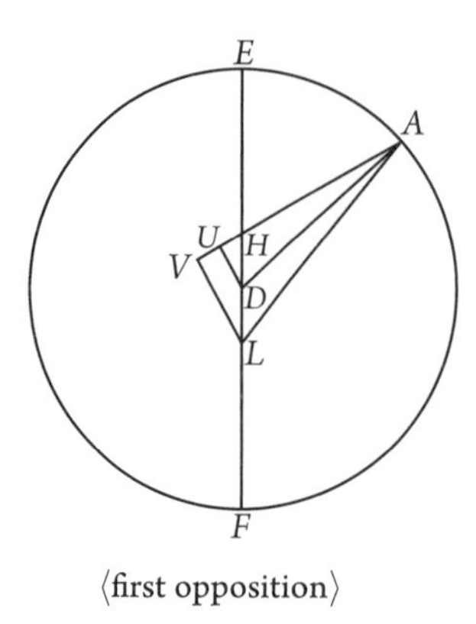

⟨first opposition⟩

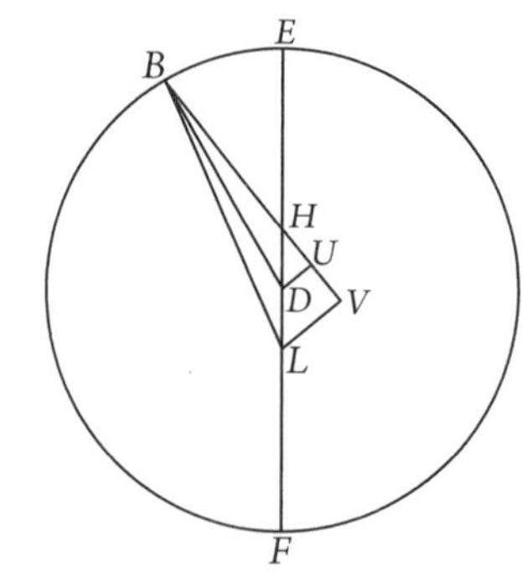

⟨second opposition⟩

And therefore angle *LAV* is 14;6 degrees of which two right-angles are 360, and 7;3 of which four right-angles are 360.

Angle *AHE* also was 41;33 of the same;
and therefore the remaining angle *ALE* of the apparent passage will be 34;30 degrees—by which the star was west of apogee in the first opposition.

Proposition 10.17 Calculation 10.15

Again, let the same diagram of the second opposition be set out. H342

Accordingly, since angle *BHE* of the mean passage of the epicycle is 40;11 degrees of which four right-angles are 360,
but both itself and angle *VHL* at its vertex are 80;22 degrees of which two-right angles are 360,
the arc on *DU* also would be 80;22 degrees of which the circle around right-angled triangle *DHU* is 360,
and the arc on *UH* would be the 99;38 degrees supplementary to the semicircle.

And therefore as to the chords subtending them:
DU is 77;26 parts of which hypotenuse *DH* is 120,
and *UH* 91;41 of the same;
so that also, *DU* also will be 3;52
and *UH*, similarly, 4;35 parts of which straight line *DH* is 6 and hypotenuse *DB* 60.

And since the square on *DU* subtracted from the square on *DB* makes the square on *BU*,
BU also in length will be 59;53 of the same.

By the same argument,[1] since *HU* is equal to *UV* and *LV* is double *DU*, then the whole *BV* will be 64;28 parts of which *LV* is 7;44. H343

And, for this reason, hypotenuse *BL* also will be 64;56 of the same.

---

[1] *By the same argument...*: following Toomer.

And therefore *LV* also will be 14;19 parts of which hypotenuse *BL* is 120, while the arc on it will be 13;42 degrees of which the circle around right-angled triangle *BLV* is 360;
so that also, angle *LBV* is 13;42 degrees of which two right-angles are 360, and 6;51 of which four right-angles are 360.

Angle *BHE* also was 40;11 of the same;
and therefore the remaining angle *ELB* of the apparent passage is 33;20 of the same.

Therefore, the star appeared eastwards of apogee in the second opposition by so many degrees.

It had been shown, for the first opposition too, being westwards of apogee by 34;30 degrees;
therefore, the whole separation, from the first opposition to the second, sums up to 67;50 degrees—in agreement with those obtained by observation {H323.5}.

Proposition 10.18

Now, in like manner, let the diagram of the third opposition also be set out.

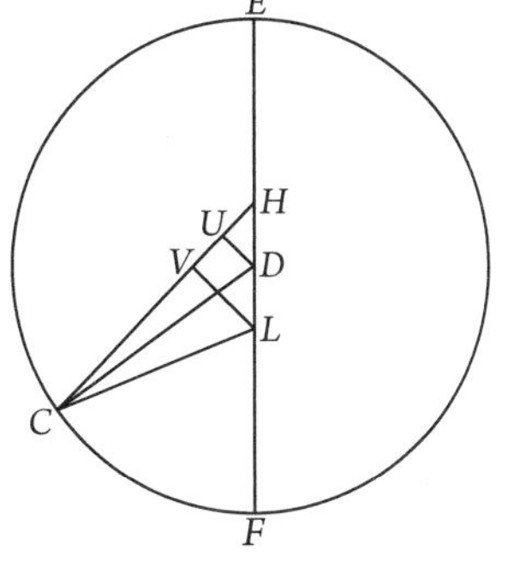

⟨third opposition⟩

Accordingly, since in this case also angle

H344

CHF, of the uniform passage of the epicycle, is 44;21 degrees of which four right-angles are 360,
and 88;42 of which two-right angles are 360,
the arc on *DU* also would be 88;42 degrees of which the circle around right-angled triangle *DHU* is 360,
while the arc on *UH* is the 91;18 degrees supplementary to the semicircle.

And therefore as to the arcs that subtend them:
DU is 83;53 parts of which hypotenuse *DH* is 120
and UH 85;49 of the same.

So that also, *DU* also will be 4;11$\frac{1}{2}$
and *UH*, similarly, 4;17 parts of which straight line *DH* is 6 and radius *DC* of the eccentric circle 60.

And since the square on *DU* subtracted from the square on *DC* makes the square on *CU*,
we will have *CU* too in length as 59;51 of the same.

Again, since *UH* also is equal to *UV*
and *LV* is double *DU*,
then we will have the remainder *VC* as 55;34 parts of which straight line LV is 8;23.

And thereby we will have hypotenuse *CL* also as 56;12 of the same.

And therefore *LV* also will be 17;55 parts of which hypotenuse *CL* is 120,
while the arc on it will be 17;10 degrees of which the circle around right-angled triangle *CLV* is 360;
so that also, angle *HCL* is 17;10 degrees of which two right-angles are 360, and 8;35 of which four right-angles are 360.

H345

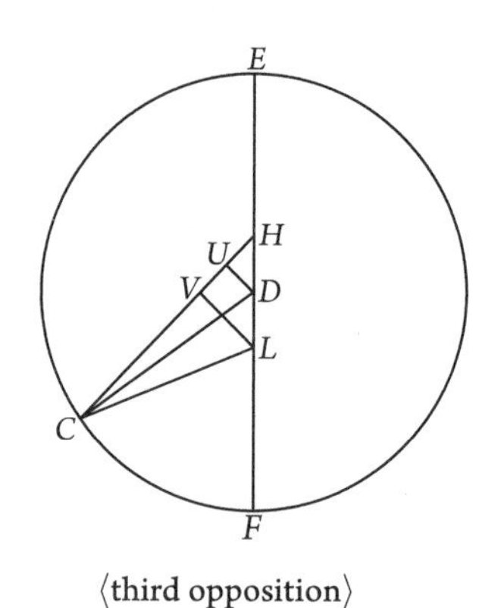

⟨third opposition⟩

Angle *CHF* also was 44;21 of the same;
and therefore the whole angle *CLF* is 52;56 of the same.

Therefore, the star appeared westwards of perigee at the third opposition by so many degrees.

Calculation 10.16

It had been shown, for the second opposition, to be eastwards of apogee by 33;20 degrees;
and therefore the remaining degrees, summed up from the second opposition, again, to the third, were found to be 93;44, in agreement with those observed in the second separation {H323.5}.

It is clear that, as well, since the star, being observed on straight line *CL* occupied the observed 2;34 degrees of Sagittarius in the third opposition {H322.13},
while angle *CLF*, at the center of the zodiac circle, was shown as 52;56 degrees of which four right-angles are 360,
then the perigee of eccentricity, the point at *F*, occupied 25;30 degrees of Capricorn,
but the apogee occupied the diametrically opposite 25;30 degrees of Cancer.

Proposition 10.19

And if we describe around center *C* epicycle *IJK* of the star Mars and produce straight line *HC*,
we will have, in the time-span of the third opposition, the mean passage of the epicycle from the apogee of the eccentric circle as 135;39 degrees,
since angle *CHF* was shown {H343.21} as the 44;21 degrees supplementary to the semicircle.

H346

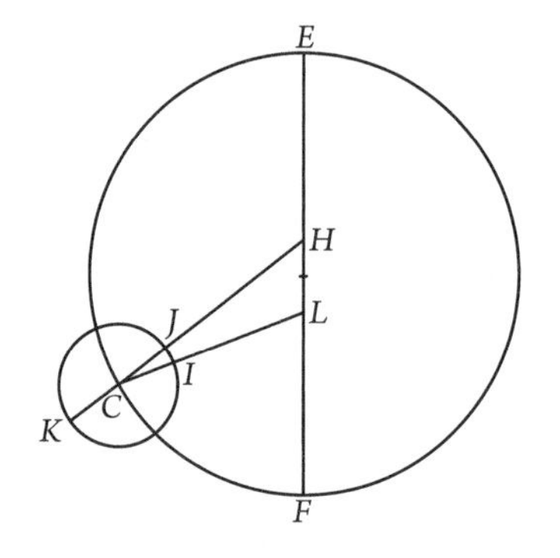

And we will have the mean passage of the star from apogee *K* of the epicycle, that is to say arc *KI*, as 171;25 degrees;

because, as angle *HCL* was shown {H345.2} to be 8;35 degrees of which four right-angles are 360 and was at the center of the epicycle,
arc *IJ* (of the star from *I* to perigee *J*) also is 8;35 of the same,
while the arc from apogee *K* to the star at *I* is the 171;25 degrees supplementary to the semicircle, as proposed.

And, along with rest, it has become clear to us that, during the time of the third opposition, that is to say in the 2nd year of Antoninus, by Egyptian reckoning Epiphi 12/13, two equinoctial hours before mid-night, the star Mars H347 in its mean passage was in its so-called longitude separated from apogee of the eccentric circle by 135;39 degrees, while in its anomaly it was separated from apogee of the epicycle by 171;25 degrees; which it was necessary to prove.

8. Demonstration of the Size of the Epicycle of the Star Mars.

Observation 10.15 Calculation 10.17

Since it follows to prove the ratio of the size of the epicycle also, for this we took an observation which we sighted three days, most nearly, after the third opposition; that is to say, in the 2nd year of Antoninus, by Egyptian reckoning Epiphi 15/16, three equinoctial hours before mid-night, since, according to the astrolabe, the 20th degree of Libra was culminating, while the sun in its mean passage occupied then 5;27 degrees of Gemini.

Accordingly, when it ⟨the astrolabe⟩ was sighted upon Spica relative to its proper position, the star Mars appeared to occupy $1\frac{3}{5}$ degrees of Sagittarius, while at the same time it also appeared separated from the center of the moon, eastwards, by the same $1\frac{3}{5}$ degrees.

And the mean passage then of the moon was around 4;20 degrees of Sagittarius, while the true passage was around 29;20 degrees of Scorpio; H348 since, in its anomaly also, it was separated from the apogee of the epicycle by 92 degrees, while its apparent passage was around the beginning of Sagittarius. So that on this ground also the star Mars then in good agreement occupied, just as it also was sighted, 1;36 degrees of Sagittarius, and was evidently separated from perigee westwards by 53;54 degrees.

In the time-span between the third opposition and this observation are contained 1;32 degrees of longitude and 1;21, most nearly, of anomaly. If we add these to the positions demonstrated according to the third opposition above {H347 ff.}, we will have at the time of this observation also the star Mars being separated from the apogee of the eccentric circle by 137;11 degrees of longitude, and from the apogee of the epicycle by 172;46 degrees of anomaly.

Proposition 10.20 Part A

Accordingly, when these values are assumed, let eccentric circle *ABC*, carrying the center of the epicycle, be around center *D* and diameter *ADC*. Upon this let the center of the zodiac circle be assumed as *E*, and the center of the greater eccentricity as *F*. And, when epicycle *GHI* is described around *B*, let *FIBG*, *EHB*, and, furthermore, *DB* be drawn through, and let *EJ* and *DK* be drawn from points *E* and *D* perpendicular to *FB*. Let the star also be assumed at point *L* of the epicycle, and, when *EL* and *BL* are joined, let *BM* be drawn perpendicular from *B* to *EL* produced.

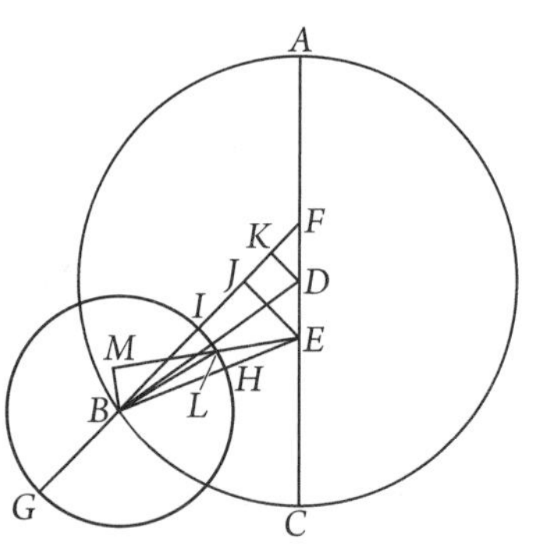

H349

Accordingly, since the star is separated from the apogee of the eccentric circle by 137;11 degrees,
so that angle *BFC* also is 42;49 degrees of which four right-angles are 360, and 85;38 of which two right-angles are 360,
the arc on *DK* also would be 85;38 degrees of which the circle around right-angled triangle *DFK* is 360,
and the arc on *FK* would be the 94;22 degrees supplementary to the semicircle.

And therefore as to the chords that subtend them:
DK will be 81;34 parts of which hypotenuse *DF* is 120,
and *FK* 88;1 of the same;
so that also, *DK* also will be 4;5
and *FK*, similarly, 4;24 parts of which the line *DF* between the centers is 6 and radius *DB* of the eccentric circle 60.

H350

And since the square on *DK* subtracted from the square on *DB* makes the square on *BK*,
straight line *BK* also will be 59;52 of the same.

Similarly, since *FK* also is equal to *KJ*
and *EJ* is double *DK*,
then the remainder *BJ* will be 55;28
and *EJ* 8;10 of the same;
and thereby hypotenuse *EB* also will be 56;4.

And therefore *EJ* also will be 17;28 parts of which straight line *EB* is 120, while the arc on it will be 16;44 degrees of which the circle around right-angled triangle *BEJ* is 360;
so that also, angle *FBE* is 16;44 degrees of which two right-angles are 360.

Proposition 10.20 Part B

Again, angle *CEM*, by which the star Mars appeared westwards of perigee *C*, is assumed as 53;54 degrees of which four right-angles are 360, and 107;48 of which two right-angles are 360,

and angle *CEB* also is 102;22 of the same,
because it is equal to angle *FBE*, shown to be 16;44 of the same, and angle *CFB*, assumed to be 85;38 of the same, taken together.

Hence, the remaining angle *BEM* also would be 5;26 of the same,

H351 while the arc on *BM* would be 5;26 degrees of which the circle around right-angled triangle *BEM* is 360;
and thereby straight line *BM* also would be 5;41 parts of which hypotenuse *EB* is 120.

And therefore *BM* also will be 2;39 parts of which EB was shown to be 56;4 and the radius of the eccentric circle 60.

Proposition 10.20 Part C

Similarly, since point *L* was separated from apogee *G* of the epicycle by 172;46 degrees,
while from perigee *I* by 7;14 degrees,
angle *IBL* also would be 7;14 degrees of which four right-angles are 360, and 14;28 of which two right-angles are 360.

Angle *IBH* also was 16;44 of the same;
and therefore the remaining angle *LBH* will be 2;16
and the whole angle *MLB* 7;42 of the same.

So that also, the arc on *MB* is 7;42 degrees of which the circle around right-angled triangle *BLM* is 360,
and straight line *BM* itself is 8;3 parts of which hypotenuse *BL* is 120.

And therefore radius *BL* of the epicycle also will be 39;30 parts, most nearly, of which straight line *BM* is 2;39 and the radius of the eccentric circle 60;
and therefore the ratio of the radius of the eccentric circle to the radius of the epicycle is 60 to 39;30; which it was proposed to find.

9. On the Correction of the Periodic Motions of the Star Mars.

H352.3

Observation 10.16
Calculation 10.18

And for the sake of correcting its periodic, mean motions, we took one of the ancient observations: it is made clear by it that in the 13th year, according to Dionysius, Aigon 25,[1] the morning star Mars seemed to have occulted the northern forehead of Scorpio. The date, then, of the observation is in the 52nd year from the death of Alexander, that is to say the 476th year from Nabonassar, by Egyptian reckoning Athur 20/21, at dawn. We find that at

[1] The Calendar of Dionysis is known only through Ptolemy's citations in *The Almagest*. The month Aigon (from αἴξ, "goat") does not correspond to any of the Egyptian months but denotes the time of year when the sun is in Capricorn. Fortunately, Ptolemy also gives the date in standard form in his next sentence.

this time the sun in its mean passage occupied 23;54 degrees of Capricorn, while the star on the northern forehead of Scorpio was observed by us to be occupying 6;3 degrees of Scorpio. So that, since again the 409 years from the observation up to the reign of Antoninus make 4;5 degrees, most nearly, of motion of the fixed stars, then the fixed star ought to have occupied 2;4 degrees of Scorpio during the time of the observation in question; and the star Mars also clearly the same 2;4 degrees.

H353 Likewise, since in our time also, that is to say at the beginning of the reign of Antoninus, the apogee of the star Mars occupied 25;30 degrees of Cancer, in the observation it ought to have occupied 21;25 degrees. And it is clear that the apparent star then was separated from apogee by 100;50 degrees, while the mean sun was separated from the same apogee by 182;29 degrees, and clearly from the perigee by 2;29 degrees.

Proposition 10.21

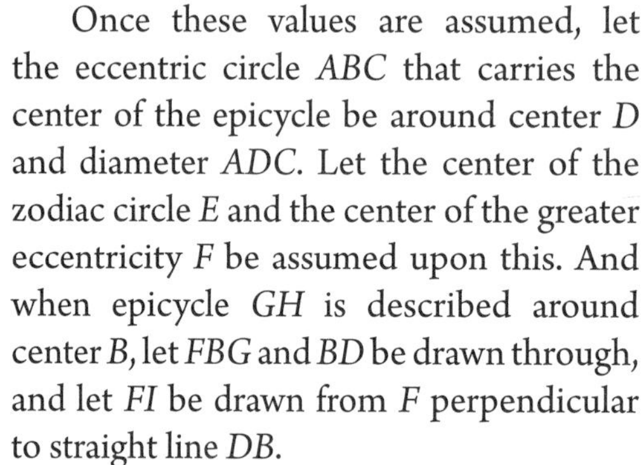

Once these values are assumed, let the eccentric circle ABC that carries the center of the epicycle be around center D and diameter ADC. Let the center of the zodiac circle E and the center of the greater eccentricity F be assumed upon this. And when epicycle GH is described around center B, let FBG and BD be drawn through, and let FI be drawn from F perpendicular to straight line DB.

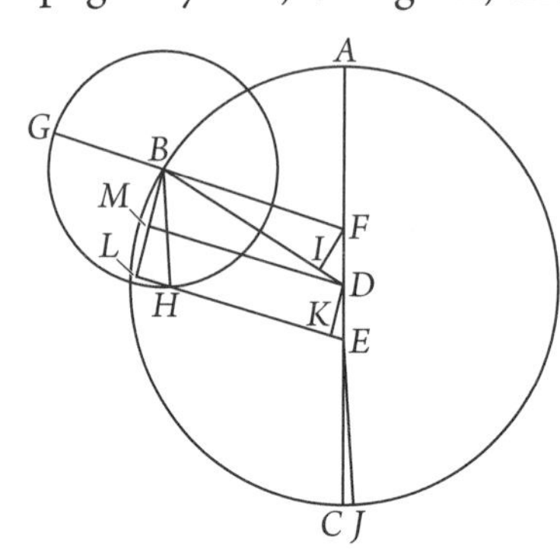

Let the star be assumed at point H of the epicycle; and when BH is joined, let EJ be drawn parallel to it from E; from what was previously demonstrated {H320.1ff.}, the mean passage of the sun will clearly be observed on this. H354

And when EH is joined, let DK and BL be drawn from D and B perpendicular to it. And, furthermore, let DM be drawn from D perpendicular to BL so that $DKLM$ is a right-angled parallelogrammic figure.

Accordingly, since angle AEH of the apparent passage of the star from apogee is 100;50 degrees of which four right angles are 360,
and angle CEJ of the mean passage of the sun is 2;29 of the same degrees, angle HEJ, that is to say angle BHE, also would be 81;39 degrees of which four right angles are 360 and 163;18 degrees of which two right angles are 360.

So that also, the arc on BL is 163;18 degrees of which the circle around right-angled triangle BHL is 360, and straight line BL itself in length is 118;43 parts of which hypotenuse BH is 120.

And therefore BL also will be 39;3 parts of which radius BH of the epicycle is 39;30 and straight line ED between the centers is 6.

Again, since angle AEH is 100;50 degrees of which four right angles are 360 and 201;40 degrees of which two right angles are 360

and thereby the angle adjacent to it, *DEK*, also is 158;20 of the same, H355
the arc on *DK* would also be 158;20 degrees of which the circle around right-angled triangle *DEK* is 360,
and straight line *DK* itself would be 117;52 parts of which hypotenuse *DE* is 120.

And therefore *DK*, that is to say *ML*, will also be 5;54 parts of which straight line *DE* is 6 and *BL* was proved to be 39;3,
and the remainder *BM* will be 33;9 parts of which radius *BD* of the eccentric circle also is 60.

And therefore *BM* also will be 66;18 parts of which hypotenuse *BD* is 120, while the arc on it will be 67;4 degrees, most nearly, of which the circle around right-angled triangle *BDM* is 360.

So that also, angle *BDM* is 67;4 degrees of which two right angles are 360,
and the whole angle *BDK* is 247;4.

Angle *EDK* also is 21;40 of the same, because angle *DEK* was shown to be 158;20 degrees.

And therefore the remaining angle *BDE* is deduced as 225;24 degrees, and angle *BDA* adjacent to it, similarly, as 134;36 degrees.

So that also, the arc on *FI* is 134;36 degrees of which the circle around right-angled triangle *DFI* is 360,
and the arc on *DI* is the 45;24 degrees supplementary to the semicircle.

And therefore as to the chords that subtend them:
H356 *FI* will be 110;32 parts of which hypotenuse *DF* is 120,
and DI is 46;18 of the same.

And therefore *FI* also will be 5;32,
DI, similarly, 2;19,
and the remaining straight line *IB* 57;41 parts of which straight line *DF* is 6 and radius *DB* of the eccentric circle is 60.

And thereby hypotenuse *BF* also will be 57;57, most nearly, of the same parts.

And therefore *FI* also will be 11;28 parts of which straight line *BF* is 120, while the arc on it will be 10;58 degrees of which the circle around right-angled triangle *BIF* is 360.

So that also, angle *FBD* is 10;58 degrees of which two right angles are 360.

Angle *BDA* also was 134;36 of the same;
and therefore the whole angle *BFA* is 145;34 of the same and 72;47 degrees of which four right angles are 360.

Therefore during the time-span of the observation in question, the mean longitudinal passage of the star, that is to say center *B* of the epicycle, was

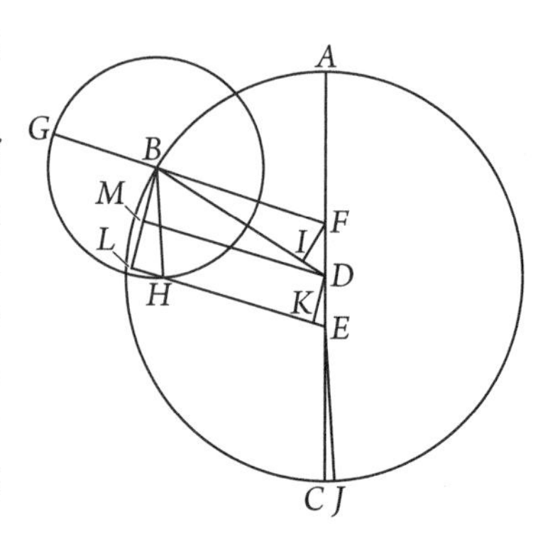

separated from apogee by 72;47 degrees and thereby occupied 4;12 degrees of Libra.

Angle *CEJ* also is assumed as 2;29 of the same degrees; this together with the two right angles of semicircle *ABC* is equal to angle *AFB* of mean longitude and angle *GBH* of anomaly (that is to say the motion of the star on the epicycle) taken together.[1] Hence we will also have the remaining angle *GBH* as 109;42 of the same degrees.

H357

Therefore at the same time of the observation the star also was separated from apogee of the epicycle by the present 109;42 degrees of anomaly; which it was proposed to find.

Calculation 10.19

At the time of the third observation, it was also shown by us as being separated (in terms of anomaly) from the apogee of the epicycle by 171;25 degrees {H346.10}.

Therefore, in the time-span between the observations that contains 410 Egyptian years and $231\frac{2}{3}$ days, most nearly, beyond 192 whole cycles, it added 61;43 degrees. This is roughly the surplus we find in the tables of its mean motions worked out by us {H232 ff.}; since the daily mean motion also was constituted by us from these—the degrees summed up from the number of cycles and the surplus divided by the days summed up from the time-span between the two observations.

10. On the Epoch of its Periodic Motions.

Calculation 10.20

Again, then, the time-span from the first year of Nabonassar, by Egyptian reckoning Thoth 1, mid-day, up to the above observation is 475 Egyptian years and $79\frac{3}{4}$ days, most nearly, and this time-span contains 180;40 degrees of surplus of longitude and 142;29 degrees of anomaly.

H358

Hence, if we subtract these values respectively from each of the positions set out according to the observations (that is to say 4;12 degrees of longitude in Libra and 109;42 degrees of anomaly), we will have for the first year of Nabonassar, by Egyptian reckoning Thoth 1, mid-day, as epoch for the periodic motions of the star Mars:
3;32 degrees of Aries in longitude
and 327;13 degrees from apogee of the epicycle in anomaly.

---

[1] This is a consequence of the relation $s = l + a$ (see the Preliminaries to Book IX). The mean longitudinal passage l from apogee is $\angle AFB$, 72;47°; while the sun's mean passage s is $\angle AEJ$ measured counterclockwise from A to J, which is 182;29°. The angle of anomaly a on the epicycle, $\angle GBH$, is the difference $s - l$ or 109;42°, as Ptolemy is about to state.

By the same considerations, since $4\frac{3}{4}$ degrees of motion of the apogees are deduced in 475 years,
and the apogee of the star Mars was around 21;25 degrees of Cancer during the observation,
it clearly will also occupy 16;40 degrees of Cancer at the time being set out for its epoch.

PRELIMINARIES TO BOOK XI

Computation of Planetary Positions

Most of Book XI is devoted to the theories of Jupiter and Saturn, which (since they are very similar in construction to the theory of Mars) are not included in the present selection. However, at the end of Book XI Ptolemy presents a set of tables that provide a short cut for computing planetary positions. Since these tables avoid the need for trigonometric computations and the extraction of square roots, they established the standard procedures henceforth used by astronomers in their calculations. In fact, this method of calculating planetary positions remained in use well into the seventeenth century. It therefore seems fitting to provide readers with the practical tools for computing genuinely Ptolemaic planetary positions.

The fundamental insight behind the construction of the tables was that planetary positions are purely angular—no radial distances are involved—and thus they can be expressed as angular additions or subtractions applied to the planet's mean position. The numerical values of both mean longitude and mean anomaly are known for the epoch (1 Thoth, year 1 of Nabonasser) and the planet's mean motion in longitude up to the desired moment is easily computed from the mean annual motion (IX.3) and the elapsed time from epoch. The "mean passage" in longitude is obtained by adding this last motion to the mean longitude at epoch, and removing full cycles. Two kinds of adjustments are then made to this mean position: one that determines the corrected position of the center of the epicycle, and the other that applies the apparent angular effect of the epicyclic motion. The former is a function solely of the mean motion in longitude in relation to the line of apsides, while the latter depends on the mean motion of anomaly, with an adjustment that derives from the mean motion of the epicycle's center.

Since Ptolemy explains how the table works without giving an account of its theoretical basis (perhaps presuming that this would be obvious), it may be helpful to describe that basis here. The operation of the table will then be outlined in a step-by-step summary at the end of these Preliminaries.

How the Tables are Constructed

In the Ptolemaic eccentric/epicycle model of planetary motion, the center of the epicycle is moved around the equant point with regular circular motion: that is, equal angles are traversed around this point in equal times. In the adjacent diagram, the epicycle's center *B* moves uniformly about the equant point *F*; however, the apparent motion of *B* as seen from the earth at *E* is nonuniform, slowest at apogee *A* and fastest at perigee *C*. If the epicycle were to move uniformly on the deferent about point *E*, it would lie on the line *EX*, which is parallel to *FB*. Thus the difference between the angle of mean motion (*AEX* or *AFB*) and the true or corrected motion *AEB* is angle *BEX* or *EBF* (which is equal to *BEX* because of the parallel lines *BF* and *XE*). This angle *EBF* is the correction that is to be applied to the mean motion to determine the position of the center of the epicycle as seen from *E*. The determination of this angle by computation requires lengthy arc and chord computations, as illustrated in Book X. However, a simple three-column table would suffice for determining the correction corresponding to any angle of mean longitude *AFB*, for each planet.

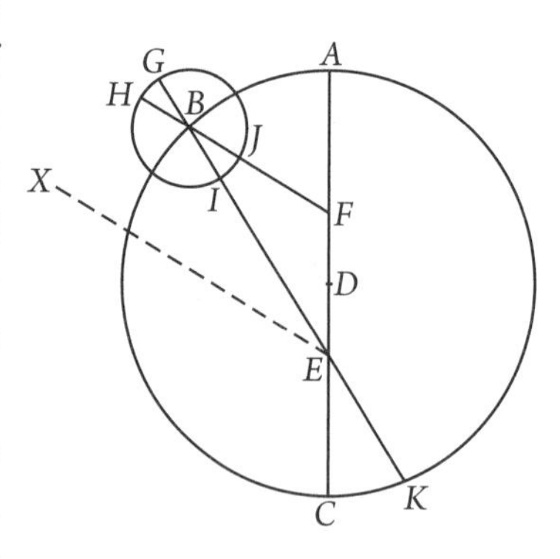

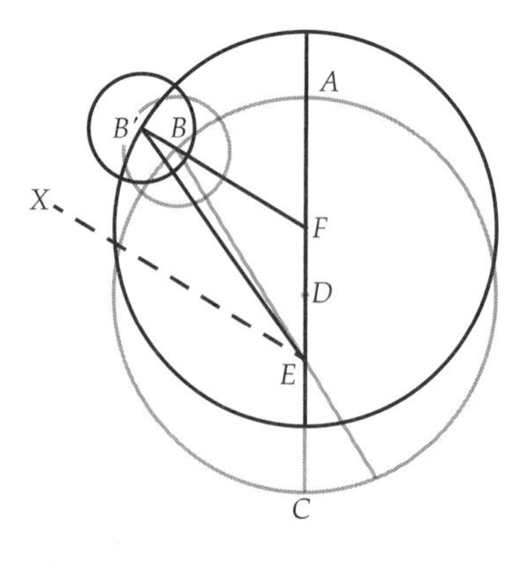

Interestingly, Ptolemy chose a somewhat less convenient approach. Instead of combining the effects of the equant (the angle about *F*) and the eccentric circle (about *D*), he first gave the correction as it would be if the eccentric circle were centered on the equant point *F*, and then gave the additional correction that would be required by moving the center of the eccentric to point *D*, halfway between the equant and the earth, *E*. The difference may be seen in the adjacent diagram. The angle about *F* remains the same, but when (as in the first correction) the eccentric (the larger black circle) is centered on *F*, the center of the epicycle at this moment is at *B′* rather than at *B*. Then, when the center of the eccentric is moved to *D*, the center of the epicycle falls at *B*, on the gray circle, resulting in another, smaller, correction, here represented by angle *B′EB*. The first correction appears in column 3 of each table, opposite the mean longitude from either column 1 or column 2, and the second appears in column 4. The total correction for the anomaly in longitude is the sum of these two partial corrections. If the mean longitude is found from column 1, the correction is subtracted from the mean value; if from column 2, it is added. This is the *adjusted longitude.*

The correction for the planet's position on the epicycle is more complicated, because it is affected not only by the planet's position on the epicycle, but also by the epicycle's distance from the earth, which changes with the longitude (for which, see the Preliminaries to Book X). The way Ptolemy does this involves three steps.

1. A mean value for the addition or subtraction due to anomaly is obtained from the known mean anomaly. This mean correction is read from column 6, opposite the value of mean anomaly in either column 1 or column 2. This gives the correction as it would be if the epicycle were at its mean distance from earth on the eccentric circle.
2. An adjustment is then applied to reflect the epicycle's actual distance from earth. First, the adjustment is calculated for the special case where the center of the epicycle is at either perigee (column 7) or apogee (column 5), whichever is closer to the planet. This adjustment will depend only on the planet's position on the epicycle (the mean anomaly) because it is calculated in each case at a fixed position on the eccentric circle. If the center of the epicycle is at perigee on the eccentric, the angular effect will be increased, and if the center of the epicycle is near apogee on the eccentric the effect will be decreased.
3. Then this adjustment is itself adjusted to reflect the angular distance of the epicycle's center from apogee or perigee. Because the change in distance on the eccentric is small, this adjustment is also small; therefore, Ptolemy provides this (in column 8) as a number that must be divided by 60 as it is used. Hence, this column is headed, "Sixtieths." Unlike the other corrections in the table, this correction is a multiplier: the number from column 8 opposite the mean longitude[1] (from either column 1 or column 2) is divided by 60 and multiplied by the result of step 2. This yields the total adjustment to the mean correction found in step 1, to be added or subtracted as indicated in columns 5, 7, and 8.

Finally the result of this three-step process is added to, or subtracted from, the adjusted longitude obtained from columns 3 and 4. The result is the planet's true or apparent removal from apogee; its "true passage" is obtained by adding this result to the position of apogee on the mid-zodiac circle, which is noted at the head of each table.

---

[1] Note that this is the angle of the epicycle on the eccentric, angle *AFB* in the above diagrams, and not the angle of anomaly on the epicycle.

How to Use the Tables

Let's take a simple example, with numbers that actually appear in a table, so as to minimize interpolation.

1. From the table for Venus, suppose the mean longitude is 294 degrees, and the mean anomaly is 125;10 degrees (this last is chosen to make the adjusted number come out conveniently).

2. Opposite 294 (column 2) the adjustment for longitude is 2;8 (column 3) + 0;2 (column 4), or 2;10. Since we are in the second semicircle, this is additive. So the adjusted longitude is 296;10. We must subtract 2;10 from the mean anomaly, which becomes 123;0.

3. Opposite 123 (column 1), the adjustment for anomaly at the mean distance on the eccentric is 44;45 (column 6). We will "write this down."

4. Opposite 294 (column 1), the "sixtieths" are 26;15 (column 8)—that is, the multiplier here is 0;26,15. This is towards apogee from the mean distance, so we use column 5. We look in this column opposite 44;45 in column 6, which we wrote down. This column 5 number is 0;57.

[If the mean longitude had been farther down in this column—at 225, say—the "sixtieths" would be 41;11, multiplier 0;41,11, and it is toward perigee from the mean distance. You can tell this because it is down below the line where the sixtieths become additive. Therefore, we use column 7. Again, we look in column 7 opposite 44;45 in column 6 and find 1;1.]

5. Returning to the column 5 number, 0;57 × 0;26,15 = 0;24,56, which rounds to 0;25. Since this stems from column 5, we subtract it from the number we "wrote down": 44;45 – 0;25 = 44;20. This is the adjustment for the anomaly. Since it is in the first semicircle, it is additive.

[In the alternative case, 1;1 × 0;41,11 = 0;41,52, which rounds to 0;42. Since this stems from column 7, we add it to the number we "wrote down": 44;45 + 0;42 = 45;27. Again, it is additive.]

6. Therefore, the total adjustment, the algebraic sum of the adjusted longitude (296;10) and the anomaly (44;20) is 340;30.

7. This is added to the position of apogee to get the apparent passage.

Book XI

H426

9. How Precise Passages are Obtained Geometrically from Periodic Motions.

The fact that, conversely,[1] when the periodic arcs, both of the eccentric circle that contains the uniform motion and of the epicycle, are given, the apparent passages of the stars are also readily obtained geometrically, will be clear to us through the same diagrams.

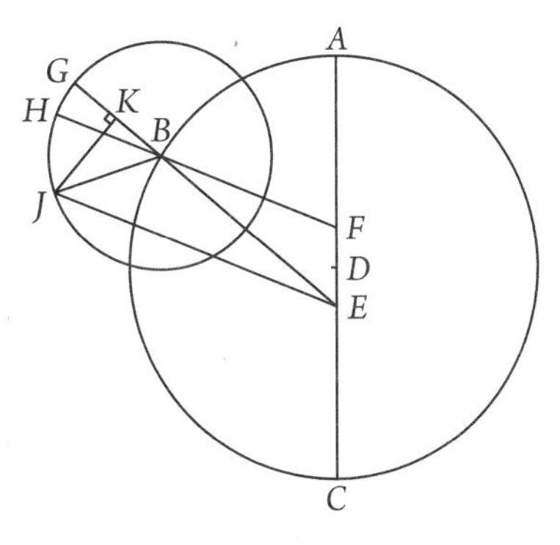

For if, in the simple diagram of the eccentric and the epicycle, we join the lines *FBH* and *EBG*, when the mean passage of longitude, that is angle *AFB*, is given, both angle *AEB* and angle *EBF*

H427

(that is, angle *GBH*) [Euc. 1.15], and furthermore the ratio of straight line *EB* to the radius of the epicycle, will also be given, on both hypotheses, on the basis of what we have previously proved. And also (for argument's sake) when the star is assumed at point *J* of the epicycle and when *EJ* and *BJ* are joined and arc *HJ* is given, if we no longer, as in the converse proof, draw a perpendicular from *B* the center of the epicycle to *EJ*, but rather from the center of the star at *J* to the straight line *EB*, as in this case *JK*, there will be given both the whole angle *GBJ*, and thereby the ratio of both *JK* and *KB* to both *BJ* [Euc. Data 40] and evidently also *EB* [Euc. Data 8], and in consequence the ratio of the whole *EBK* to *KJ* [Euc. Data 6, 8]; so

---

[1] In the earlier parts of Book XI, Ptolemy showed how the arcs on the eccentric and the epicycle, as well as all the parameters of the planets, can be deduced from the given apparent passages together with the mean motions. He now begins to show how the direction of the argument can be reversed.

that, since angle *KEJ* is given [Euc. Data 41],[1] there has been obtained for us angle *AEJ* [Euc. Data 3], which contains the apparent distance of the star from apogee.

10. Treatment of Table-Construction of the Anomalies.

But in order that we not always calculate the apparent passages geometrically—this sort of approach, though it alone renders the matter at hand precise, turns out to be rather difficult for ease of enquiries—we have elaborated, in a way that is as useful as possible and at the same time nearest to precision, a table of each of the five stars containing their particular, combined anomalies, in order that, through the periodic motions themselves, given from their respective apogees, we also readily calculate their apparent passages in each case. We have, then, arranged each of the tables in 45 rows, again for the sake of a common measure, and eight columns. The first two columns will contain the numerical values of the mean passages,[2] just as in the table of the sun {III.6} and that of the moon [V.8, not included in these selections]; in the first column the 180 degrees from apogee are arranged from above downward, and in the second column the remaining 180 degrees of the semicircle from the bottom upward, so that the numerical value of 180 degrees be arranged in both bottom rows. Their augmentation occurs in the first 15 top rows through 6 degree ⟨intervals⟩ while in the 30 rows below them through 3 degree ⟨intervals⟩; since, of the segments of anomaly, the differences at apogees do not differ from one another to a great extent, while at perigees they change more quickly.

H428

H429

Of the following two columns, the third will contain addition-subtractions, of the mean longitudinal passage due to the greater eccentricity,[3] corresponding to the numeric values of their respective rows, but these addition-subtractions are taken in a simple manner, as if the center of the epicycle were carried on the same eccentric circle itself that contains the uniform motion.[4] The fourth column

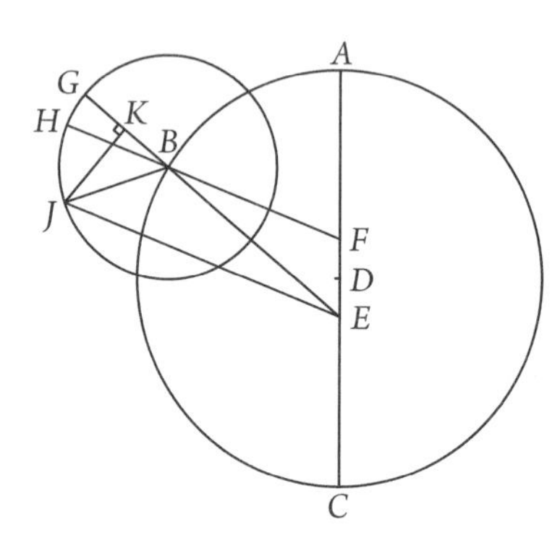

[1] When two sides of the right triangle *EKJ* are given, the angle *KEJ* is also given.

[2] "Mean passages" is plural because these columns are used to enter both the mean longitudinal motion and the mean anomaly.

[3] *The greater eccentricity*: *EF*, the eccentricity of the equant.

[4] That is, the third column shows the addition-subtraction that would result if the eccentric circle bearing the epicycle were centered on *F* rather than on *D*. The fourth column then

will contain the resultant differences of the addition-subtractions due to the center of the epicycle's not being carried on the previously mentioned circle but rather on the other circle.[1] The manner in which each of these is obtained geometrically, together and separately, has become clear to us through the numerous theorems we have previously set out.

In this case, then, it was appropriate, in terms of proper order, to bring such a determination of the zodiacal anomaly into view and thereby set it out in two columns, but in terms of utility itself, just one column combined from the addition-subtraction of both of these suffices.

Of the next three columns, each will contain the addition-subtractions due to the epicycle, again taken simply ⟨at first,⟩ and ⟨then⟩ as if the apogees or perigees in them are observed in relation to the distance from our H430 sight;[2] and the manner of this kind of demonstration has become readily understood by us through the previously set-forth theorems. The middle one, then, of these three columns (and the sixth from the first) will contain the addition-subtractions that are derived in accordance with the ratios of the mean distances. The fifth will contain the differences, occurring on the same segments, of the addition-subtractions at the greatest distance compared to those at the mean distance; and the seventh will contain the difference of the addition-subtractions at the least distance compared with those at the middle distance. For we have shown that

where the radius of the epicycle of the star Saturn is 6;30 parts
(for it would be well in what follows to begin from the upper stars)
and of the star Jupiter is 11;30 parts,
and of the star Mars 29;30 parts,
and of the star Venus 43;10 parts,
and of the star Mercury 22;30 parts,

in these units the mean distance of all them is 60 parts: that is to say, the distance observed in relation to the radius of the eccentric circle carrying the epicycle; and the greatest distance ⟨of the center of the epicycle from earth⟩, taken in relation to the center of the zodiac circle, is

63;25 for Saturn,
62;45 for Jupiter,
H431 66 for Mars,
61;15 for Venus, and
69 for Mercury.

---

adds another correction that results from moving that center back down to *D*. See the Preliminaries to Book XI for a fuller explanation.

[1] *On the other circle*: circle *ABC*, with center *D*.

[2] As will be explained below, column 6 presents the simple or mean addition-subtraction, and columns 5 and 7 contain the corrections to be applied at apogee and perigee, respectively.

The least distance is, in just the same way,

56;35 for Saturn,
57;15 for Jupiter,
54 for Mars,
58;45 for Venus, and
55;34 for Mercury.[1]

The remaining column, the eighth, has been arranged by us to obtain the corresponding parts of the differences we are setting out, whenever the epicycles of the stars do not happen to be at the mean, greatest or least distances, but in the passages between these.[2] And we have also arranged the calculation of such a correction with respect only to the greatest addition-subtractions that at each of the intermediate distances are produced by tangents to the epicycle from our sight; as the value of the differences in the case of particular segments of the epicycle differs in no significant way compared to the differences in the case of the greatest addition-subtractions.[3]

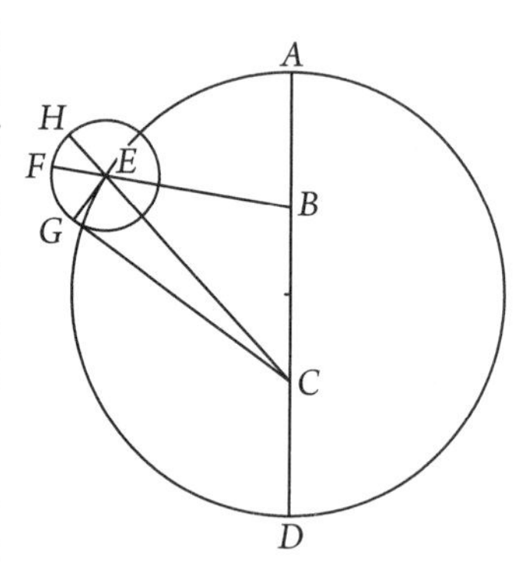

And for the sake of both what we are stating becoming clearer and rendering the method of the applications ⟨of the calculations⟩ itself clear, let there be set out the straight line *ABCD* through both centers, of the zodiac circle and of the eccentric circle that contains the uniform motion of the epicycle, and let *C* be assumed as the center of the zodiac circle, and *B* as the center of the uniform motion of the epicycle, and with *BEF* extended, let epicycle *FG* be described around center *E*, and let straight line *CG* be drawn from *C* tangent to it, and let *CE* and perpendicular *EG* [Euc. 3.18] be joined.

H432

[1] These distances are carried over from the theories of the individual planets. The distances for Mercury are asymmetrical about the mean because of its more complicated theory, which includes an extra circle.

[2] This column is the only one that functions as a multiplier rather than as an additive or subtractive adjustment: it is therefore simply a dimensionless number and not an angle or a length. At apogee and perigee its value is 1, and it becomes 0 at the mean distance. This provides a modification of the adjustment for greatest or least distance (columns 5 or 7, respectively), keeping it at its maximum value at apogee or perigee and reducing it to zero at mean distance, where the value from column 6 is used without adjustment. To make the table less cluttered, Ptolemy expresses the numbers in column 8 as "sixtieths," so that, for example, 0;59,53 is entered as "59 53" at the top of column 8. This multiplier is applied to column 5 when it is subtractive, and to column 7 when additive.

[3] The multiplier is treated as a linear factor, although the relations involved may not be strictly linear, because any departure from linearity is insignificant.

And for each of the five stars, let the center of the epicycle be assumed, by way of example, as being uniformly[1] separated from apogee of eccentricity by 30 degrees. Since (so as not to make the calculation lengthy by proving the same things) it has been shown through many ⟨theorems⟩ in what precedes in the case of both the hypothesis of Mercury and of the rest of the stars, that when angle *ABE* is given, the ratio of *CE* to the radius of the epicycle (that is, *GE*) is also given, and this is yielded through each calculation of angle *ABE* assumed as thirty of those degrees where four right angles are 360:

H433 in the case of Saturn the ratio of 63;2 to 6;30,
of Jupiter the ratio of 62;26 to 11;30,
for Mars the ratio of 65;24 to 39;30,
for Venus the ratio of 61;26 to 43;10, and
for Mercury the ratio of 66;35 to 23;30;

then we will have the angle *ECG*, which contains the greatest addition-subtraction at that time due to the epicycle: where four right angles are 360, in these units

for Saturn 5;55½,
for Jupiter 10;36½,
for Mars 37;9,
for Venus 44;56½, and
for Mercury 19;45.

And there are derived also the greatest addition-subtractions at the mean distances according to the ratios we set out just shortly above, respectively (for the present order of the stars, in order that we not be redundant):

6;13 and 11;3 and 41;10 and 46;0 and 22;2 degrees;

those at the greatest distances

5;53 and 10;34 and 36;45 and 44;48 and 19;2;

those at the least distances:

H434 6;36 and 11;35 and 47;1 and 47;17 and 23;53,

so that those at the greatest distances differ from those at the mean distances by

0;20 and 0;29 and 4;25 and 1;12 and 3;0 degrees,

and those at the least distances by

0;23 and 0;32 and 5;51 and 1;17 and 1;51 degrees.

Since, then, the addition-subtractions of the distances we are seeking are smaller than those at the mean distances and differ from them by

0;17½ and 0;26½ and 4;1 and 1;3½ and 2;17 ⟨respectively⟩,

and the sixtieths of the whole differences that are being set out of the mean distances in relation to the greatest differences are,

---

[1] *Uniformly*: in its mean motion; that is, angle *ABE* is 30°.

for Saturn 52;30 and
for Jupiter 54;50 and
for Mars 54;34 and
for Venus 52;55 and
for Mercury 45;40,

we have entered these many 60ths in the eighth columns of each table in the row containing the numerical value of 30 degrees of periodic longitude.

In the case of distances that have greater addition-subtractions compared with those at the mean distances, we have resolved their resultant differences in the same way into 60ths, in relation, however, to the whole differences of the addition-subtractions at the least distances and no longer of those at the greatest distances. And by calculating the resultant 60ths of the whole differences in the same way, in the case of the rest of the positions, through 6 degree ⟨intervals⟩, we set them out alongside the respective numerical values of the value of the differences that, as we said, is perceptibly the same even if the passages of the stars do not occur at the greatest addition-subtractions of the epicycle, but even at all the rest of its parts. And the exposition of the five tables is as follows. H435

11. Tables of the Longitudinal Adjustment of the Five Planets.

H436

Saturn: Apogee 14;10 degrees of Scorpio.

1	*2*	*3*		*4*		*5*		*6*		*7*		*8*	
common numbers		*additions-subtractions of longitude*		*difference of additions-subtractions*		*difference of subtraction*		*additions-subtractions of anomaly*		*difference of addition*		*60ths of subtraction*	
6	354	0	37	0	2	0	2	0	36	0	2	60	0
12	348	1	13	0	4	0	4	1	11	0	4	58	30
18	342	1	49	0	6	0	5	1	45	0	7	57	0
24	336	2	23	0	8	0	7	2	18	0	9	55	30
30	330	2	57	0	9	0	8	2	50	0	11	52	30
36	324	3	29	0	10	0	10	3	20	0	13	49	30
42	318	3	59	0	11	0	11	3	49	0	15	46	30
48	312	4	28	0	11	0	12	4	17	0	17	43	30
54	306	4	55	0	10	0	14	4	42	0	19	39	0
60	300	5	20	0	9	0	15	5	4	0	20	34	30
66	294	5	42	0	8	0	17	5	25	0	20	30	0
72	288	6	0	0	7	0	18	5	42	0	21	24	0
78	282	6	14	0	5	0	18	5	55	0	21	18	0
84	276	6	24	0	3	0	19	6	5	0	22	12	0
90	270	6	30	0	1	0	19	6	12	0	22	4	30
93	267	6	31	0	0	0	20	6	12	0	23	0	45
				subtraction								*of addition*	
96	264	6	32	0	2	0	20	6	13	0	23	2	32
99	261	6	31	0	3	0	20	6	12	0	24	5	51
102	258	6	30	0	4	0	21	6	12	0	24	9	8
105	255	6	27	0	5	0	21	6	9	0	24	11	45
108	252	6	23	0	6	0	20	6	5	0	25	14	21
111	249	6	19	0	7	0	20	6	0	0	25	16	58
114	246	6	14	0	8	0	20	5	55	0	24	19	31
117	243	6	7	0	9	0	19	5	48	0	24	22	11
120	240	5	59	0	10	0	19	5	40	0	23	24	47
123	237	5	50	0	10	0	19	5	31	0	23	27	24
126	234	5	39	0	11	0	18	5	21	0	22	30	0
129	231	5	27	0	11	0	18	5	10	0	22	32	37
132	228	5	14	0	12	0	17	4	58	0	21	35	13
135	225	5	0	0	12	0	17	4	45	0	20	37	50
138	222	4	45	0	12	0	16	4	31	0	19	40	26
141	219	4	29	0	12	0	15	4	16	0	18	43	3
144	216	4	12	0	12	0	14	4	0	0	17	45	39
147	213	3	54	0	12	0	14	3	43	0	15	47	37
150	210	3	35	0	11	0	12	3	25	0	14	49	34
153	207	3	16	0	11	0	11	3	7	0	13	51	32
156	204	2	56	0	10	0	10	2	48	0	12	53	29
159	201	2	36	0	9	0	9	2	29	0	11	54	49
162	198	2	15	0	8	0	7	2	9	0	10	56	6
165	195	1	53	0	7	0	6	1	48	0	8	57	24
168	192	1	31	0	6	0	5	1	27	0	7	58	42
171	189	1	9	0	5	0	5	1	6	0	5	59	21
174	186	0	47	0	3	0	4	0	45	0	4	60	0
177	183	0	24	0	2	0	2	0	23	0	2	60	0
180	180	0	0	0	0	0	0	0	0	0	0	60	0

H437

Jupiter: Apogee 2;9 degrees of Virgo.

H438

1	2	3		4		5		6		7		8	
common numbers		*additions-subtractions of longitude*		*difference of additions-subtractions*		*difference of subtraction*		*additions-subtractions of anomaly*		*difference of addition*		*60ths of subtraction*	
6	354	0	30	0	1	0	2	0	58	0	2	60	0
12	348	1	0	0	2	0	5	1	56	0	5	58	58
18	342	1	30	0	3	0	7	2	52	0	7	57	56
24	336	1	58	0	4	0	9	3	48	0	9	56	54
30	330	2	26	0	5	0	11	4	42	0	11	54	50
36	324	2	52	0	6	0	13	5	34	0	13	51	43
42	318	3	17	0	7	0	15	6	25	0	15	47	35
48	312	3	40	0	7	0	17	7	12	0	18	43	27
54	306	4	1	0	7	0	19	7	57	0	20	39	19
60	300	4	20	0	6	0	21	8	37	0	22	35	8
66	294	4	37	0	5	0	23	9	14	0	24	28	58
72	288	4	51	0	4	0	24	9	46	0	26	22	45
78	282	5	2	0	3	0	25	10	13	0	28	17	35
84	276	5	9	0	2	0	26	10	35	0	30	11	23
90	270	5	14	0	1	0	26	10	51	0	31	4	40
93	267	5	15	0	0	0	27	10	57	0	31	1	8
				subtraction								*of addition*	
96	264	5	16	0	1	0	27	11	0	0	32	1	52
99	261	5	15	0	1	0	27	11	2	0	32	5	9
102	258	5	14	0	2	0	28	11	3	0	32	8	26
105	255	5	12	0	2	0	28	11	1	0	33	11	43
108	252	5	9	0	3	0	29	10	59	0	33	15	0
111	249	5	5	0	4	0	29	10	53	0	33	17	49
114	246	5	0	0	5	0	30	10	45	0	34	20	37
117	243	4	54	0	5	0	30	10	35	0	34	23	26
120	240	4	47	0	6	0	30	10	24	0	34	26	15
123	237	4	39	0	6	0	29	10	10	0	33	29	4
126	234	4	30	0	7	0	29	9	54	0	33	31	52
129	231	4	20	0	7	0	28	9	36	0	32	34	41
132	228	4	9	0	8	0	28	9	16	0	32	37	30
135	225	3	58	0	8	0	27	8	54	0	31	40	19
138	222	3	46	0	8	0	26	8	30	0	30	43	7
141	219	3	33	0	8	0	25	8	4	0	28	45	28
144	216	3	20	0	7	0	23	7	36	0	26	47	49
147	213	3	6	0	7	0	22	7	6	0	25	49	42
150	210	2	51	0	6	0	21	6	34	0	23	51	31
153	207	2	36	0	6	0	19	6	0	0	21	52	58
156	204	2	20	0	5	0	17	5	24	0	19	54	22
159	201	2	4	0	5	0	15	4	47	0	17	55	47
162	198	1	47	0	4	0	13	4	9	0	15	57	11
165	195	1	30	0	3	0	11	3	29	0	13	57	40
168	192	1	13	0	2	0	9	2	49	0	10	58	13
171	189	0	55	0	2	0	7	2	7	0	8	58	40
174	186	0	37	0	1	0	5	1	25	0	5	59	4
177	183	0	18	0	1	0	3	0	43	0	3	59	32
180	180	0	0	0	0	0	0	0	0	0	0	60	0

H439

H440

Mars: Apogee 16;40 degrees of Cancer.

1	*2*	*3*		*4*		*5*		*6*		*7*		*8*	
common numbers		*additions-subtractions of longitude*		*difference of additions-subtractions*		*difference of subtraction*		*additions-subtractions of anomaly*		*difference of addition*		*60ths of subtraction*	
6	354	1	0	0	5	0	8	2	24	0	9	59	53
12	348	2	0	0	10	0	16	4	46	0	18	58	59
18	342	2	58	0	15	0	24	7	8	0	28	57	51
24	336	3	56	0	20	0	33	9	30	0	37	56	36
30	330	4	52	0	24	0	42	11	51	0	46	54	34
36	324	5	46	0	27	0	51	14	11	0	56	52	11
42	318	6	39	0	28	1	0	16	29	1	6	49	28
48	312	7	28	0	29	1	9	18	46	1	16	46	17
54	306	8	14	0	28	1	18	21	0	1	28	42	38
60	300	8	57	0	27	1	27	23	13	1	40	38	8
66	294	9	36	0	24	1	37	25	22	1	53	33	26
72	288	10	9	0	20	1	49	27	29	2	6	28	20
78	282	10	38	0	15	2	1	29	32	2	19	22	47
84	276	11	2	0	10	2	14	31	30	2	33	16	33
90	270	11	19	0	4	2	28	33	22	2	45	10	5
93	267	11	25	0	0	2	35	34	15	2	57	6	34
				subtraction									
96	264	11	29	0	4	2	42	35	6	3	6	3	3
												of addition	
99	261	11	32	0	8	2	49	35	56	3	15	0	5
102	258	11	32	0	12	2	56	36	43	3	25	3	13
105	255	11	31	0	16	3	4	37	27	3	36	6	1
108	252	11	28	0	19	3	13	38	9	3	47	8	49
111	249	11	22	0	22	3	22	38	48	3	58	11	44
114	246	11	14	0	25	3	32	39	24	4	9	14	38
117	243	11	5	0	28	3	43	39	56	4	21	17	33
120	240	10	53	0	31	3	54	40	23	4	35	20	27
123	237	10	39	0	33	4	4	40	44	4	50	23	35
126	234	10	23	0	35	4	14	40	59	5	5	26	42
129	231	10	4	0	37	4	24	41	7	5	21	29	31
132	228	9	44	0	39	4	35	41	9	5	37	32	20
135	225	9	21	0	40	4	45	41	2	5	55	35	9
138	222	8	55	0	41	4	56	40	45	6	14	37	58
141	219	8	27	0	41	5	7	40	16	6	34	40	35
144	216	7	59	0	41	5	18	39	37	6	53	43	12
147	213	7	27	0	41	5	28	38	40	7	12	45	26
150	210	6	54	0	38	5	34	37	25	7	30	47	39
153	207	6	19	0	36	5	38	35	52	7	45	49	50
156	204	5	41	0	33	5	38	33	53	7	58	52	1
159	201	5	3	0	30	5	34	31	30	8	3	53	47
162	198	4	22	0	27	5	18	28	35	7	58	55	32
165	195	3	41	0	23	4	52	25	3	7	47	56	44
168	192	2	58	0	19	4	18	21	0	7	6	57	55
171	189	2	14	0	15	3	32	16	25	5	59	58	49
174	186	1	30	0	10	2	27	11	15	4	26	59	43
177	183	0	45	0	5	1	16	5	45	2	20	59	52
180	180	0	0	0	0	0	0	0	0	0	0	60	0

H441

Venus: apogee 16;10 degrees of Taurus.

H442

1 2 *common numbers*		3 *additions-subtractions of longitude*		4 *difference of additions-subtractions*		5 *difference of subtraction*		6 *additions-subtractions of anomaly*		7 *differenence of addition*		8 *60ths of subtraction*	
6	354	0	14	0	1	0	1	2	31	0	2	59	10
12	348	0	28	0	1	0	3	5	1	0	4	57	55
18	342	0	42	0	1	0	5	7	31	0	6	56	40
24	336	0	56	0	2	0	7	10	1	0	8	55	0
30	330	1	9	0	2	0	9	12	30	0	10	52	55
36	324	1	21	0	2	0	11	14	58	0	12	49	35
42	318	1	32	0	3	0	13	17	25	0	14	45	50
48	312	1	43	0	3	0	15	19	51	0	16	42	5
54	306	1	53	0	3	0	18	22	15	0	18	37	5
60	300	2	1	0	2	0	20	24	38	0	20	31	40
66	294	2	8	0	2	0	22	26	57	0	23	26	15
72	288	2	14	0	2	0	24	29	14	0	25	20	25
78	282	2	18	0	1	0	27	31	27	0	28	14	35
84	276	2	21	0	1	0	29	33	38	0	30	8	20
90	270	2	23	0	1	0	31	35	44	0	33	1	40
				subtraction								*of addition*	
93	267	2	23	0	0	0	33	36	40	0	36	1	31
96	264	2	23	0	1	0	35	37	43	0	38	4	42
99	261	2	22	0	1	0	38	38	40	0	40	7	39
102	258	2	21	0	1	0	40	39	35	0	43	10	35
105	255	2	20	0	1	0	42	40	29	0	45	13	32
108	252	2	18	0	1	0	45	41	20	0	47	16	28
111	249	2	16	0	1	0	47	42	9	0	50	19	25
114	246	2	13	0	2	0	49	42	54	0	52	22	21
117	243	2	10	0	2	0	52	43	35	0	55	25	18
120	240	2	6	0	2	0	54	44	12	0	58	28	14
123	237	2	2	0	2	0	57	44	45	1	1	31	0
126	234	1	58	0	2	1	0	45	14	1	4	33	44
129	231	1	54	0	2	1	3	45	36	1	8	36	18
132	228	1	49	0	3	1	6	45	51	1	11	38	50
135	225	1	44	0	3	1	10	45	59	1	14	41	11
138	222	1	39	0	3	1	14	45	57	1	18	43	32
141	219	1	33	0	3	1	19	45	45	1	22	45	42
144	216	1	27	0	2	1	24	45	20	1	27	47	51
147	213	1	21	0	2	1	29	44	40	1	32	49	37
150	210	1	14	0	2	1	33	43	39	1	38	51	23
153	207	1	7	0	2	1	37	42	18	1	43	52	46
156	204	1	0	0	2	1	39	40	28	1	48	54	8
159	201	0	53	0	2	1	41	38	7	1	51	55	18
162	198	0	46	0	1	1	42	35	7	1	52	56	26
165	195	0	39	0	1	1	38	31	24	1	50	57	28
168	192	0	32	0	1	1	31	26	46	1	43	58	26
171	189	0	24	0	1	1	19	21	15	1	27	59	1
174	186	0	16	0	1	0	58	14	47	1	5	59	36
177	183	0	8	0	1	0	31	7	38	0	35	59	58
180	180	0	0	0	0	0	0	0	0	0	0	60	0

H443

H444

Mercury: Apogee 1;10 degrees of Libra.

1 common numbers	2	3 additions-subtractions of longitude		4 difference of additions-subtractions		5 difference of subtraction		6 additions-subtractions of anomaly		7 difference of addition		8 60ths of subtraction	
6	354	0	18	0	1	0	10	1	38	0	5	59	20
12	348	0	34	0	2	0	20	3	16	0	11	57	20
18	342	0	51	0	4	0	29	4	53	0	17	54	40
24	336	1	7	0	5	0	39	6	29	0	23	50	40
30	330	1	22	0	5	0	49	8	4	0	28	45	40
36	324	1	37	0	4	0	59	9	36	0	34	39	40
42	318	1	51	0	4	1	8	11	6	0	40	33	0
48	312	2	4	0	3	1	18	12	33	0	45	25	40
54	306	2	15	0	1	1	28	13	58	0	50	18	0
60	300	2	25	0	0	1	39	15	18	0	56	10	20
				addition									
66	294	2	34	0	2	1	49	16	33	1	4	2	20
												of addition	
72	288	2	41	0	4	1	59	17	43	1	11	9	14
78	282	2	46	0	6	2	9	18	47	1	17	20	0
84	276	2	50	0	7	2	19	19	44	1	23	29	44
90	270	2	52	0	9	2	29	20	33	1	29	39	28
93	267	2	52	0	10	2	34	20	54	1	32	43	31
96	264	2	52	0	10	2	39	21	14	1	35	47	34
99	261	2	51	0	11	2	44	21	29	1	38	50	0
102	258	2	50	0	10	2	48	21	42	1	41	52	26
105	255	2	48	0	10	2	53	21	52	1	44	54	52
108	252	2	46	0	10	2	58	21	59	1	46	57	18
111	249	2	44	0	9	3	2	22	2	1	49	58	23
114	246	2	41	0	9	3	4	22	1	1	52	59	28
117	243	2	37	0	9	3	6	21	56	1	55	59	44
120	240	2	33	0	8	3	8	21	47	1	57	60	0
123	237	2	28	0	7	3	9	21	33	1	59	59	44
126	234	2	23	0	7	3	10	21	15	2	0	59	23
129	231	2	18	0	6	3	12	20	53	2	0	58	39
132	228	2	12	0	6	3	12	20	25	2	1	57	50
135	225	2	6	0	5	3	9	19	50	2	1	56	46
138	222	2	0	0	4	3	6	19	10	2	0	55	41
141	219	1	53	0	4	3	2	18	24	2	0	54	3
144	216	1	46	0	3	2	57	17	32	1	58	52	26
147	213	1	38	0	3	2	51	16	35	1	53	50	48
150	210	1	30	0	2	2	42	15	31	1	47	49	11
153	207	1	22	0	2	2	32	14	20	1	41	47	34
156	204	1	13	0	2	2	21	13	3	1	34	45	57
159	201	1	5	0	1	2	9	11	41	1	26	44	36
162	198	0	56	0	1	1	55	10	13	1	17	43	15
165	195	0	46	0	1	1	38	8	40	1	7	42	26
168	192	0	38	0	0	1	19	7	1	0	56	41	37
171	189	0	28	0	0	1	1	5	19	0	43	40	48
174	186	0	19	0	0	0	42	3	35	0	28	40	0
177	183	0	9	0	0	0	21	1	48	0	14	39	44
180	180	0	0	0	0	0	0	0	0	0	0	39	28

H445

Note on the tables: The numbers are translated from the Greek numerals of Heiberg's edition, with Toomer's corrections (*Almagest* p. 548 note 55).

12. On the Longitudinal Calculation of the Five Planets.

H446

Whenever, then, we wish to discover the apparent passages of each of the five stars from their periodic motions of longitude and of anomaly through our treatment of the foregoing, we will set out the reasoning behind the calculation that is one and the same for the five planets as follows.

For, from the tables of mean motion, by combining the uniform positions of longitude and of anomaly near the required time, beyond whole cycles, we will first enter into the respective table of anomaly of the star the degrees from the apogee of the eccentric circle at that time up to the mean longitudinal passage; and if the numerical value of longitude set out occurs in the first column, we will subtract what is set out alongside the numeric value in the third column of the longitudinal adjustment, together with the combined addition-subtraction of 60ths in the fourth column, from the degrees of longitude, and we will add it to the degrees of anomaly; but if it occurs in the second column, we will add ⟨it⟩ to the degrees of longitude and subtract ⟨it⟩ from the degrees of anomaly, in order that we have both passages adjusted.[1]

H447

And next by again applying the adjusted numerical value of anomaly from apogee to the first two columns, we will write down[2] the addition-subtraction for the mean distance, which is set alongside it in the sixth column. Also, applying the original, previously introduced, number of uniform longitude to the same numerical values ⟨in column 1 or 2⟩, if the numeric value occurs in the first rows, nearer to apogee than the mean distance—this is made clear

[1] The adjustment for the longitude is subtractive in the first semicircle and additive in the second, as is clear from the geometry. But further, in order to give the correct angles for the triangle *CEG*, which determines the adjustment in the anomaly, the geometry also requires the anomaly to be measured from the true apogee of the epicycle (angle *HEG* rather than angle *FEG*). Since Ptolemy's tables of mean anomaly are constructed with reference to the "mean apogee" (that is, the point on the epicycle at greatest distance from the equant point, *F* in the diagram), a correction to the mean anomaly must be made. This correction is equal to the adjustment in longitude: it is the arc *FH*, which is numerically equal to the angle *CEB*, which is the total adjustment in longitude.

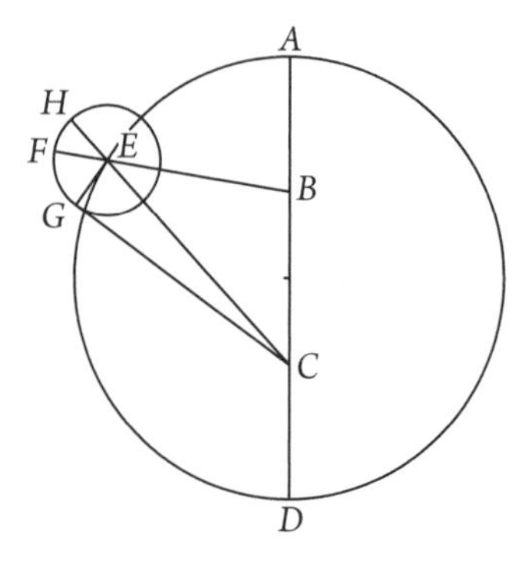

[2] We need to "write it down" because it is going to be needed for determining the adjustment that reflects the actual distance of the epicycle from the vision. Column 6 only contains the adjustment for the mean distance, and so must not be added or subtracted just yet.

from the 60ths in the eighth column[1]—then however many the 60ths set alongside it in the same eighth column are, multiplying this quantity ⟨of sixtieths⟩ by the difference in the fifth column, in the ⟨same⟩ row as the mean addition-subtraction just written down, we will subtract the resultants from what we wrote down. If, on the other hand, the numeric value of the stated longitude is in the lower rows, closer to perigee than the mean distance, ⟨then⟩ however many 60ths similarly set alongside it in the eighth column there are, ⟨by⟩ multiplying this quantity ⟨of 60ths⟩ by the difference set out alongside the mean addition-subtraction written down in the seventh column of least distance, multiplying this quantity ⟨of sixtieths⟩ by the difference in the seventh column, in the row set out alongside the mean addition-subtraction we ⟨just⟩ wrote down, in the seventh column of the least distance, we will add the resultants to those which we wrote down ⟨before⟩.[2] And as for the combined degrees of adjusted addition-subtraction, if the adjusted value of anomaly occurs in the first column, we will add them to the degrees of adjusted longitude, but if in the second, we will subtract them. And starting counting the combined numerical value of degrees from the apogee of the star at that time, we will arrive at its apparent passage.

H448

[1] The first part of column 8, which is labeled "subtractive," includes all the rows that are closer to apogee. The remaining rows are closer to perigee. If the number of the mean longitudinal motion, read from either column 1 or column 2, corresponds to one of the subtractive sixtieths in column 8, then this number (divided by 60) is multiplied by the number in column 5, that is adjacent to the number from column 6 that we "wrote down" above, and the product is subtracted from the number obtained from column 6.

[2] Ptolemy is evidently having a hard time describing the correct procedure. For a clearer explanation, see "How to Use the Tables" on p. 202.

PRELIMINARIES TO BOOK XII

Alternative Hypotheses

In Book XII Ptolemy will put aside consideration of the zodiacal anomaly to focus exclusively on the heliacal anomaly—that apparent irregularity which, for all planets, involves a sequence of *station, retrogradation,* followed once more by *station.* Recall that up to now Ptolemy has employed an epicyclic hypothesis for the heliacal anomaly of the planets. As he will recount at H450 below, the specific features of that hypothesis are that

> the epicycle makes its longitudinal passage eastwards through the signs of the zodiac around the circle concentric with the zodiac circle, while the star on the epicycle makes its passage of anomaly around its center eastwards through the arc of apogee...

That is to say, the motion of the planet on the epicycle has the *same direction* (namely, eastward) as does the motion of the epicycle on the deferent.[1]

As we saw in Book X, Ptolemy has two versions of the same-direction epicyclic hypothesis; they differ only in the relation between the epicycle and the mean sun. In the form employed for Venus (and also for Mercury, though with additional complications), the center of the epicycle moves longitudinally with average speed equal to that of the mean sun; thus the center of the epicycle never departs very far from the mean sun, and the planet's elongations are limited by the size of the epicycle. But Mars, Jupiter, and Saturn are capable of any elongation from the mean sun; therefore in the version of the epicyclic hypothesis applied to them, the speed of the epicycle is necessarily *not equal* to that of the mean sun.

In the present Book, however, he recognizes an alternative hypothesis for Mars, Jupiter, and Saturn. At H451 he will call this alternative an *eccentric* hypothesis, in which

---

[1] In principle, there can also be an epicycle hypothesis in which the motion of the star on the epicycle is *opposite* to that of the epicycle on the deferent. As we saw in Book III, Ptolemy applied such a case to the Sun and showed that it was there equivalent to a simple eccentric; see Prop. 3.5. For the planets, however, Ptolemy has employed only the *same-direction* epicycle hypothesis.

> the center of the eccentric circle is carried around the center of the zodiac circle eastwards through the signs of the zodiac with the same speed as the sun,[1] while the star on the eccentric circle is carried around its center westwards through the signs of the zodiac with the same speed as the passage of anomaly...

Note that in contrast to the stationary eccentric Ptolemy earlier employed for the sun's anomaly, the eccentric here being introduced is *moving*: for he describes its center as being "carried eastward about the ecliptic's center." The following sketches compare the two hypotheses.

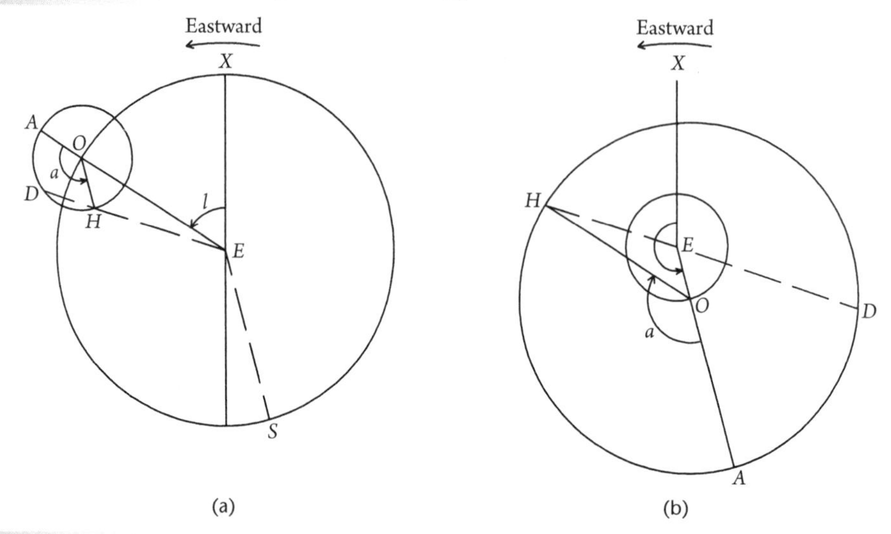

(a) (b)

Sketch (a) represents the epicyclic hypothesis for Mars; *E* is our eye, located at the center of the mid-zodiac circle.[2] We are to suppose that the epicycle's center *O* was initially at *X* but now occupies the position shown. And it is clear that when the epicycle's center was at *X*, its apogee *A* would have been in line with *E* and *X*; suppose also that Mars was then at apogee *A* but now occupies the position *H* shown. Angle *XEO* equals *l*, the longitudinal motion of the epicycle; angle *AOH* equals *a*, the anomalistic motion of the planet. Finally, point *S* represents the position of the mean sun. If *S*, too, originally coincided with *X*, then angle *XES* (figuring eastward) represents the motion of the mean sun. As Ptolemy noted in Book X, that angle is the sum of longitudinal motion *l* and anomalistic motion *a*; and thus *ES* will always be parallel to *OH*.

Sketch (b) represents the alternative hypothesis. Again point *E* is our eye at the center of the ecliptic; but now *O* is the center of the eccentric, and thus

---

[1] *the sun*: Here, the *mean sun*.

[2] As noted at the outset of this section, Ptolemy is ignoring the zodiacal anomaly; thus he does not distinguish between the center of the ecliptic, the center of the deferent, and the equant center.

EO is the eccentricity while *A* is apogee. With the same initial assumptions as before, the eccentric's center *O* has been carried eastward through angle *XEO*, while the star has moved westward on the eccentric through angle *AOH*.

But Ptolemy's characterization of the alternative hypothesis in the passage quoted above specifies that angle *XEO* in sketch (b) is equal to the angle of the mean sun; which means that *EA* in sketch (b) must be parallel to both *ES* and *OH* in sketch (a). Furthermore, he specifies that angle *AOH* in sketch (b) is equal to the angle of anomalistic motion, *a*; it follows then that *OH* in sketch (b) must be parallel to *EO* in sketch (a). Then the triangles *HEO* in both sketches are similar, so that *EH* in sketch (b) will be parallel to *EH* in sketch (a). Thus the two hypotheses yield identical longitudes for the star *H* and are therefore equivalent.

In light of the foregoing equivalence, we may choose to consider the moving eccentric in sketch (b) as being simply *an epicycle that is larger than its deferent.* Doing so, moreover, bestows a satisfying unity on the planetary hypotheses, since the epicycles of *all* planets may then be said to move longitudinally with the speed of the mean sun.

Furthermore, the epicycles of Venus and Mercury will be smaller than their deferents, while the epicycles of Mars, Jupiter, and Saturn will be larger than their deferents. If we take the deferents of all the planets as being equal in diameter to the stationary eccentric of the sun,[1] Mercury and Venus would revolve in circles that are smaller than the sun's, while Mars, Jupiter, and Saturn would revolve in circles that are larger than the sun's—thereby suggesting a quite literal meaning for the conventional distinction between "inner" and "outer" planets. Ptolemy, however, nowhere explicitly takes this step.

The Station Criterion

For the small-epicycle hypothesis Ptolemy will show that a planet at *H* manifests the appearance of station if (in sketch a)

$$\tfrac{1}{2}DH : EH :: \text{epicycle's speed} : \text{star's speed},$$

that is, if

$$\tfrac{1}{2}DH : EH :: l : a.$$

Such is the *station criterion* for the small-epicycle hypothesis. For the large-epicycle hypothesis, on the other hand, he will prove the station criterion to be (in sketch b)

[1] Recall that, while Ptolemy has established the *ratio* of diameters of each planet's epicycle to its deferent, he has no way to determine the diameters of epicycle or deferent absolutely; thus nothing prevents us from supposing an arbitrary diameter for the deferents of all the planets.

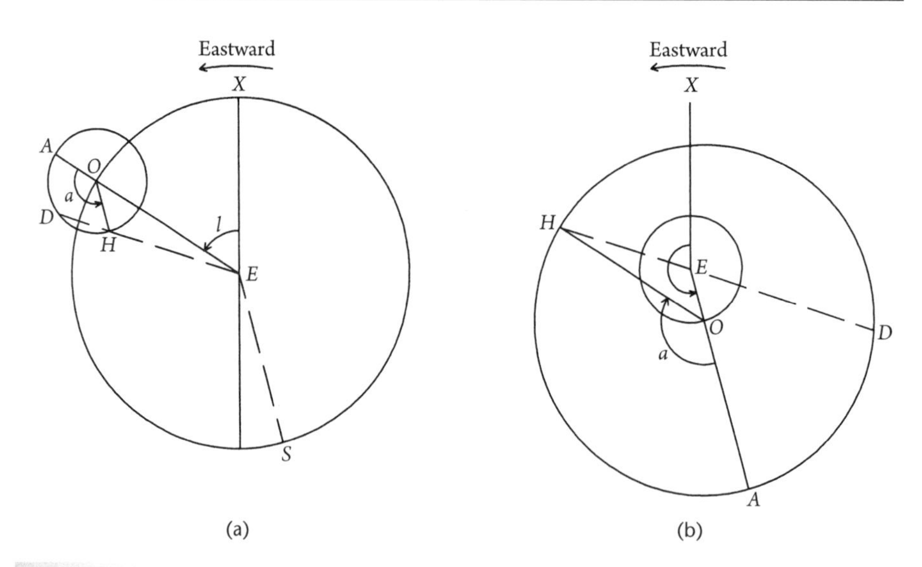

(a) (b)

$$\tfrac{1}{2}DH : EH :: (l+a) : a.$$

But since we proved the equivalence of the hypotheses, it is easy to see that the two criteria are themselves equivalent. For we already noted that angle *XES* in sketch (a) equals the sum $l+a$, while *EA* in sketch (b) is parallel to *ES* in sketch (a); so that angle *XEA* in sketch (b) is also equal to $l+a$. Thus $l+a$ represents the longitudinal motion of the large epicycle. And a represents the star's motion on the epicycle in both hypotheses. Thus for the large-epicycle as well as the small-epicycle hypothesis, the station criterion will be

$$\tfrac{1}{2}DH : EH :: \text{epicycle's speed} : \text{star's speed}.$$

Book XII

1. On the Preliminaries to the Retrogradations.

H450.3

Now that these things have been demonstrated, it would follow to investigate the least and greatest retrogradations for each of the five planets and to demonstrate as well that the magnitudes that result from the above hypotheses are in agreement, to the greatest extent possible, with those gotten from observations.

Now,[1] for a determination of this kind, all the other mathematicians and especially Apollonius of Perga prove first for the heliacal anomaly alone[2] that:

Proposition 12.1

If the heliacal anomaly arises through the epicyclic hypothesis (when the epicycle makes its longitudinal passage eastwards through the signs of the zodiac around the circle concentric with the zodiac circle, while the star on the epicycle makes its passage of anomaly around its center eastwards through the arc of apogee), and if a certain straight line from our sight is drawn through, cutting the epicycle in such a way that half of the segment it cuts off within the epicycle has to the straight line from our sight up to its

H451

point of intersection with the epicyle at perigee the same ratio that the speed of the epicycle has to the speed of the star, then the point that is produced by the line drawn through in this way at the arc near perigee of the epicycle delimits its forward motions and retrogradations;[3] so that the star, when at this point, makes an appearance of a station.[4] And they prove first that:

---

[1] Propositions 12.1 and 12.2 which follow are enunciations, without proof, of the two criteria for station that Ptolemy will prove in Propositions 12.3–8. In these latter proofs, Ptolemy shows that, to some extent at least, the epicyclic hypothesis he has used heretofore to account for the planets' heliacal anomalies is equivalent to an alternative hypothesis introduced here for the first time. The Preliminaries to Book XII describe this alternative and give a more explicit proof of the equivalence.

[2] *the helical anomaly alone*: In this chapter Ptolemy treats the heliacal anomaly (of which retrogradation is a central feature) in isolation from the zodiacal anomaly.

[3] *the point ... delimits its forward motions and retrogradations*: that is, the point in question separates the planet's progressions from its retrogradations.

[4] Note that Ptolemy is conceiving the station point as a boundary between progressive motion and retrogradation. This understanding is essential to the proofs that follow.

Proposition 12.2 If the heliacal anomaly results through the eccentrical hypothesis[1] (since this hypothesis can be viable only for the three stars that make every elongation from the sun),[2] when the center of the eccentric circle is carried around the center of the zodiac circle eastwards through the signs of the zodiac with the same speed as the sun, while the star on the eccentric circle is carried around its center westwards through the signs of the zodiac with the same speed as the passage of anomaly, and if a certain straight line on the eccentric circle is drawn through the center of the zodiac circle (that is to say, our sight) such that half of its whole length has to the lesser of the segments produced by the sight the same ratio that the speed of the eccentric circle has to the speed of the star, then the star, when at that point at which the straight line cuts the arc near perigee of the eccentric circle, will make the appearance of stations.

We too will set forth the preceding in a way that is no less useful for its brevity, by employing for both hypotheses a common and mixed demonstration to indicate their agreement in these ratios and their similarity. H452

Proposition 12.3 For let epicycle *ABCD* be around center *E*, and let its diameter be *AEC*, produced to center *F* of the mid-zodiac circle, that is to say, our sight. And when equal arcs *CG* and *CH* are cut off on each side of perigee *C*, let *FGB* and *FHD* be drawn through from *F* through points *G* and *H*. And let *DG* and *BH* be joined, cutting each other at point *I*, which clearly will fall upon diameter *AC* [Euc. 1.4, Euc. 3.7]. We say, first, that straight line *AI* is to straight line *IC* as straight line *AF* is to straight line *FC*.[3]

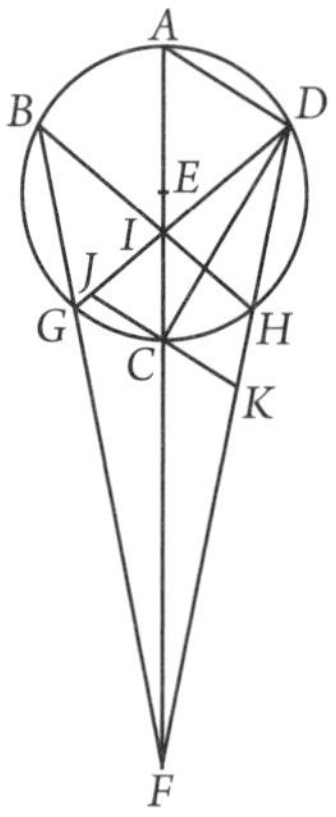

For let *AD* and *DC* be joined, and through *C* let *JCK* be drawn parallel to *AD*, clearly at right angles to *DC* [Euc. 1.29], since angle *ADC* is right [Euc. 3.21]. H453

Accordingly, since angle *CDG* is equal to angle *CDH* [Euc. 3.27], straight line *CJ* also is equal to straight line *CK* [Euc. 1.26];
and therefore *AD* has the same ratio to each of them.

But, *AF* is to *FC* as *AD* is to *CK*,
and *AI* is to *IC* as *AD* is to *JC* [Euc. 6.4];
and therefore *AI* is to *IC* as *AF* is to *FC*.

---

[1] *the eccentrical hypothesis*: Ptolemy here refers to a new kind of eccentric hypothesis, described in the remainder of the paragraph. It may also be named the "large-epicycle" hypothesis, as suggested in the Preliminaries to Book XII.

[2] *since this hypothesis can be viable only for the three stars that make every elongation from the sun*: Contrary to Ptolemy's assertion, it can be proved that with suitable adjustment of the rates of uniform motion, the "large-epicycle" hypothesis can apply to Venus as well.

[3] Alternatively stated, *AC* is divided internally and externally in the same ratio by points *I* and *F*, respectively; see Appendix 3.

Therefore, if we conceive of epicycle *ABCD* (on the eccentrical hypothesis) as the eccentric circle itself, then point *I* will be the center of the zodiac circle, and diameter *AC* will be divided by it into the same ratio on the epicyclic hypothesis; since we showed that *AF* (the greatest distance) has to *FC* (the least distance) on the epicycle, the same ratio than *AI* (the greatest distance) has to *IC* (the least distance) on the eccentric circle also.

Proposition 12.4

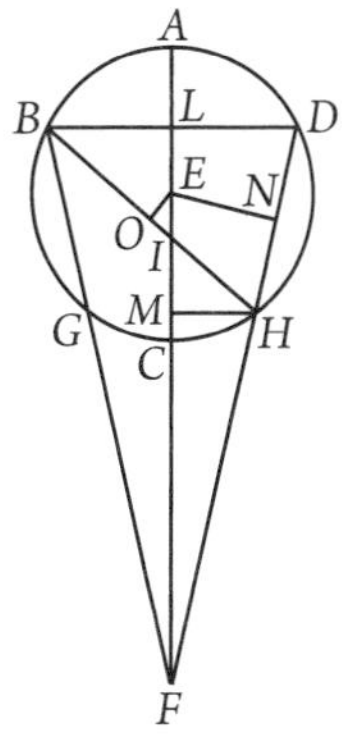

We say that straight line *BI* also has to straight line *IH* the same ratio that straight line *DF* has to straight line *FH*.

For, in the like diagram,[1] let straight line *BLD* be joined, clearly at right angles to diameter *AC* [Euc. 1.4], and from *H* let *HM* be drawn parallel to it.

H454

Accordingly, since *BL* is equal to *LD*,
therefore each of them has the same ratio to *MH*.

But *DF* is to *FH* as *LD* is to *MH*,
and *BI* is to *IH* as *BL* is to *MH* [Euc. 6.4];
and therefore *BI* is to *IH* as *DF* is to *FH*.

And therefore, *componendo*, *BH* is to *HI* as *DF*, *FH* is to *FH* [Euc. 5.18];
and *separando*,[2] when *EN* and *EO* are drawn perpendicular, *OH* is to *IH* as *NF* is to *FH* [Euc. 3.3, Euc. 5.15, Euc. 5.17].

And, furthermore, *separando*, *OI* is to *IH* as *NH* is to *FH*[3] [Euc. 5.17].

Therefore, if, on the epicyclic hypothesis, *DF* is so drawn through that *NH* has to *FH* the ratio which the speed of the epicycle has to the speed of the star, then straight line *OI* will have to straight line *IH* the same ratio on the eccentrical hypothesis also. And the reason for not using here this separated ratio for the stations (that is to say the ratio of *OI* to *IH*) but the unseparated ratio (that is to say, that of *OH* to *IH*) is that the speed of the epicycle has to the speed of the star the ratio which the longitudinal passage alone has to the passage of anomaly, and the speed of the eccentric circle has to the speed of the

H455

[1] As the Preliminaries to Book XII explains, equivalence of the small- and large-epicycle hypotheses requires the planet to move eastward on the small epicycle but westward on the large epicycle or moving eccentric. Ptolemy's composite proof thus requires that we look at the figure simultaneously from the front (for one hypothesis) and from the back (for the alternative hypothesis).

[2] *separando*: Rather, *dividendo*—taking the halves of the antecedents in relation to the consequents.

[3] The ratio *NH*: *FH* meets the definition Ptolemy gives at H250 to make *H* a station point on the small epicycle, and *OH*: *IH* fulfills the corresponding definition for the large epicycle or moving eccentric. Ptolemy has not yet proved that the planet at *H* does indeed make the appearance of a station. The present theorem only locates the point described by Ptolemy's original enunciations and proves that it has the same significance for both hypotheses.

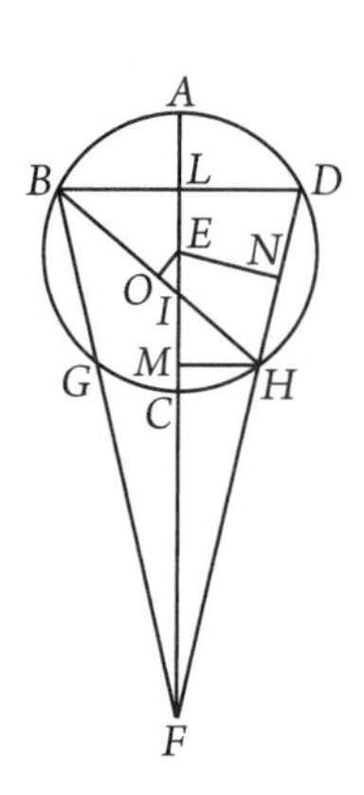

star the ratio which the mean passage of the sun (that is to say, the longitudinal passage and the passage of anomaly of the star combined)[1] has to the passage of anomaly. So that, for example, for the star Mars the ratio of the speed of the epicycle to the speed of the star is that of 42, most nearly, to 37 {H215.1-5}; for the ratio of the longitudinal passage to the passage of anomaly was shown by us to be this quantity, most nearly {IX.3}. And for this reason *NH* also has this ratio to *HF*. And the ratio of the speed of the eccentric circle to the speed of the star is that of both added together, 79, to 37; that is to say, the ratio of *OH* to *HI* taken *componendo*, since the ratio *separando* of *OI* to *IH* was the same as the ratio of *NH* to *HF*, that is to say, of 42 to 37. H456

And let these matters be given this much of a preliminary examination. It remains to be shown that, when the straight lines, taken on each of the hypotheses, are divided in such a ratio,[2] points *G* and *H* will contain the appearances of stations,[3] and that it is necessary that arc *GCH* is in retrogradation, while the remainder is in forward motion.[4] Hence, Apollonius presupposes a little lemma as follows:

Lemma 12.1

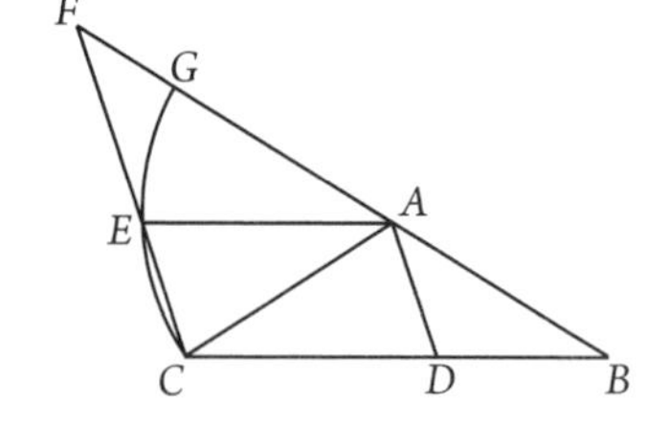

In triangle *ABC* with side *BC* greater than *AC*, if *CD*, not smaller than *AC*, is cut off, then *CD* will have a greater ratio to *BD* than angle *ABC* has to angle *BCA*.

He proves this in the following way. H457

For let parallelogram *ADCE* be completed, he says, and let *BA* and *CE* produced meet at point *F*.

Since *AE* is not less than *AC*,
therefore the circle described with center *A* and distance *AE* will go either through *C* or beyond *C*.

Now, let *GEC* be described through *C*.

And since triangle *AEF* is greater than sector *AEG*,
but triangle *AEC* is smaller than sector *AEC*,

[1] In the case of Mars, Jupiter, and Saturn, the motion of the mean sun equals the sum of the regular longitudinal motion of the planet and its motion in anomaly ($s = l + a$).

[2] *such a ratio*: that is, such that *OI* : *IH* or *NH* : *FH* :: *speed of the epicycle* : *speed of the star on the epicycle* for the small-epicycle hypothesis, or *OH* : *IH* or *NF* : *FH*:: *speed of the epicycle* : *speed of the star on the epicycle* for the large-epicycle or moving eccentric hypothesis.

[3] *points G and H will contain the appearances of stations*: as asserted in Propositions 12.1–2.

[4] *forward motion*: that is, eastward through the Zodiac signs.

triangle *AEF* has to triangle *AEC* a greater ratio than sector *AEG* has to sector *AEC*.

But, angle *EAF* is to angle *EAC* as sector *AEG* is to sector *AEC*,
and base *FE* is to base *EC* as triangle *AEF* is to triangle *AEC*;
therefore *FE* has to *EC* a greater ratio than angle *FAE* has to angle *EAC* [Euc. 6.1].

But *CD* is to *DB* [Euc. 6.2] as *FE* is to *EC*, and angle *FAE* is equal to angle *ABC*, and angle *EAC* is equal to angle *BCA* [Euc. 1.29];
and therefore *CD* has to *DB* a greater ratio than angle *ABC* has to angle *ACB*.

H458 And it is clear that the ratio will be even much greater, if *CD*, that is to say, *AE*, is not assumed equal to *AC* but greater.

Proposition 12.5

When this lemma is presupposed, let epicycle *ABCD* be around center *E* and diameter *AEC*. Let this be produced to point *F* (our sight) such that *EC* has to *CF* a greater ratio than the speed of the epicycle has to the speed of the star.[1]

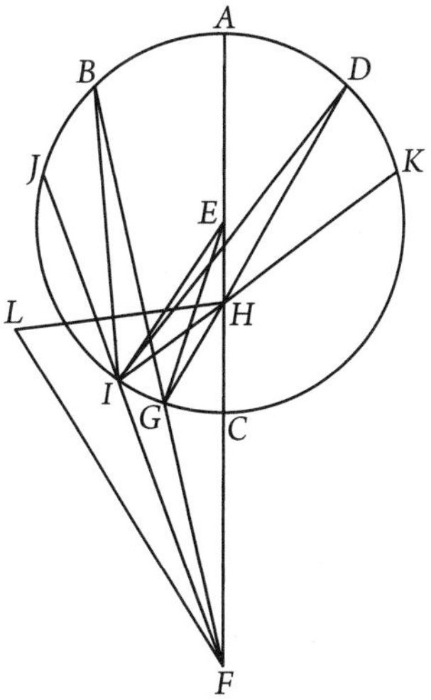

Therefore it is possible [Euc. 3.8] to draw straight line *FGB* through such that half of *BG* has to *GF* the ratio which the speed of the epicycle has to the speed of the star.

And if, through what was previously proved, we cut off arc *AD* equal to arc *AB* and join *DHG*, point *H*, on the eccentrical hypothesis, will be conceived of as our sight,
while half of *DG* will have to *HG* the same ratio which the speed of the eccentric circle has to the speed of the star {H455.21}.

H459 Now, we say that the star, when at point *G*, will on each of the hypotheses make an appearance of a station, and whatever size arc we cut off on either side of *G*, we will find the arc cut off near apogee in forward motion, while the arc near perigee in retrogradation.

For let chance arc *IG* be cut off near apogee first, and let *FIJ* and *IHK* be drawn through, and let *BI*, *DI*, and, furthermore, *EI* and *EG* be joined.

Therefore, since, in triangle *BIF*, *BG* is greater than *BI* [Euc. 3.15], *BG* has to *GF* a greater ratio than angle *GFI* has to angle *GBI* {H456.10ff};
so that half of *BG* has to *GF* a greater ratio than angle *GFI* has to the angle double angle *IBG*, that is to say, angle *IEG* [Euc. 3.20].

And the ratio of the speed of the epicycle to the speed of the star is that of half *BG* to *GF*;

---

[1] *such that EC has to CF a greater ratio than the speed of the epicycle has to the speed of the star*: Ptolemy will argue in Proposition 12.8 below that it is only under this condition that stations and retrogradation will appear.

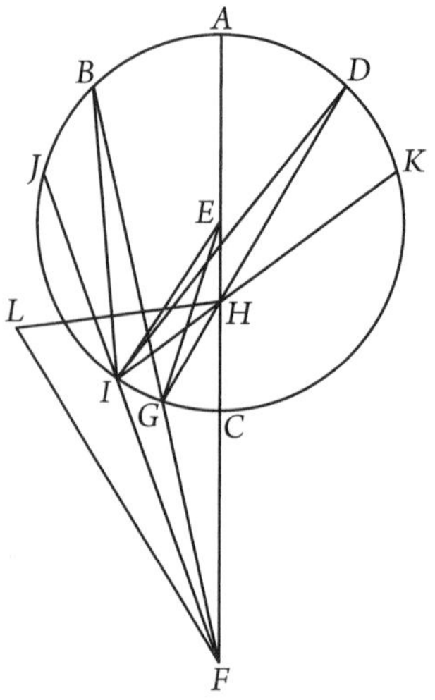

therefore angle *GFI* has to angle *IEG* a lesser ratio than the speed of the epicycle has to the speed of the star.

Therefore the angle with the same ratio to angle *IEG* that the speed of the epicycle has to the speed of the star is greater than angle *GFI*.

Now, let it be angle *GFL*.[1]

The center of the epicycle, then, has moved in the opposite direction[2] a passage equal to the distance from *FG* to *FL*[3] in the amount of time in which the star moves through arc *IG* of the epicycle.

Hence, it is clear that in the same amount of time arc *IG* of the epicycle has carried the star angle *GFI* westwards a lesser angle (at our sight) than that which the epicycle itself has moved it eastwards; that is to say, a lesser angle than *GFL*.

So that the star has moved angle *IFL* eastwards.

H460

Proposition 12.6

Similarly, if we make a calculation for the eccentric circle,
since *BG* has to *GF* a greater ratio than angle *GFI* has to angle *GBI*,
therefore *componendo* *BF* also has to *FG* a greater ratio than angle *BIJ* [Euc. 1.32] has to angle *GBI*.

But *DH* is to *HG* as *BF* is to *FG* [Prop. 12.4],
and angle *BIJ* is equal to angle *DIK*,
while angle *GBI* is equal to angle *GDI* [Euc. 3.27];
therefore, *DH* also has a greater ratio to *HG* than angle *DIK* has to angle *GDI*.

H461

So that, *componendo*, *DG* has a greater ratio to *GH* than angle *GHI* has to angle *GDI*;
and therefore, *separando*,[4] half of *DG* has to *GH* a greater ratio than angle *GHI* has to the angle double angle *GDI*, that is to say, angle *GEI* [Euc. 3.20].

And the ratio of the speed of the eccentric circle to the speed of the star is that of half of *DG* to *HG*;
therefore angle *GHI* has a lesser ratio to angle *GEI* than the speed of the eccentric circle has to the speed of the star.

Therefore the angle with the same ratio to angle *GEI* that the speed of the eccentric circle has to the speed of the star is greater than angle *GHI*.

---

[1] *angle GFL*: Note that there is no particular significance to the point *L*.

[2] *the opposite direction*: that is, eastward, since as it approaches perigee the star on the epicycle is moving westward through the zodiac.

[3] *a passage equal to the distance from FG to FL*: that is, through an angle equal to angle *GFL*.

[4] *separando*: Rather, *dividendo*, as previously in Proposition 12.4.

Now, again, let it be angle *GHL*.[1]

Accordingly, since in the same time the star itself, being moved arc *IG*, has moved angle *IEG* westwards,[2] but, due to the motion of the eccentric circle itself, has been moved angle *GHL*, greater than angle *IHG*, eastwards, it is clear that in this way also the star will appear to have moved angle *IHL* forward.

H462 It is easily seen at once that, through the same considerations, the opposite[3] will be proved also, if, in the same diagram, we assume that half of *JI* has to *IF* the same ratio that the speed of the epicycle has to the speed of the star, so that half of *KI* also has to *HI* the same ratio that the speed of the eccentric circle has to the speed of the star, and if we conceive arc *IG* as cut off from straight line *JF* towards the perigee. Proposition 12.7

For when *JG* is joined and produces triangle *JFG*, in which *FI* is cut off greater than *FG* [Euc. 3.8],

JI will have a lesser ratio to *IF* than angle *GFI* has to angle *GJI* {Lem. 12.1}.

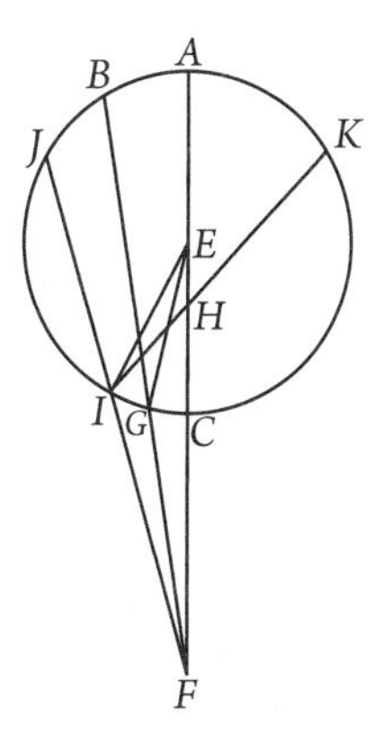

So that half of *JI* has a lesser ratio to *IF* than angle *GFI* has to the angle double angle *GJI*, that is to say, angle *IEG* [Euc. 3.20], in the opposite way to what was previously proved.[4]

And through the same considerations it will be deduced that,[5] inversely, angle *IEG* has a lesser ratio to angle *GFI* than the speed of the star has to the speed of the epicycle, and a lesser ratio to angle *GHI* than the speed of the star has to the speed of the eccentric circle. H463

So that, since the angle with the same ratio is greater than angle *IEG*, the retrograde motion also is produced greater than the forward motion.

It is clear that, for those distances at which *EC* does not have a greater ratio to *CF* than the ratio which the speed of the epicycle has to the speed of the star, neither will it be possible to draw another straight line through in the same ratio, nor will the star appear to be in station or retrogradation. Proposition 12.8

---

[1] Here too, as in Proposition 12.5, there is nothing particularly significant about the position of point *L*.

[2] Note that Ptolemy has here reversed his usual practice and treats westward motion as counter-clockwise. It can be seen as clockwise if the figure is imagined as if seen from behind the page.

[3] *the opposite*: that is, that the apparent motion will be retrograde.

[4] In Proposition 12.5, it was shown that angle *GFI* : angle *GEI* < epicycle's speed : star's speed. Under the assumptions of this theorem, epicycle's speed : star's speed < angle *GFI* : angle *GEI*.

[5] Since whenever $A : B < C : D$ it follows that $B : A > D : C$; compare Euc. V. Def. 7.

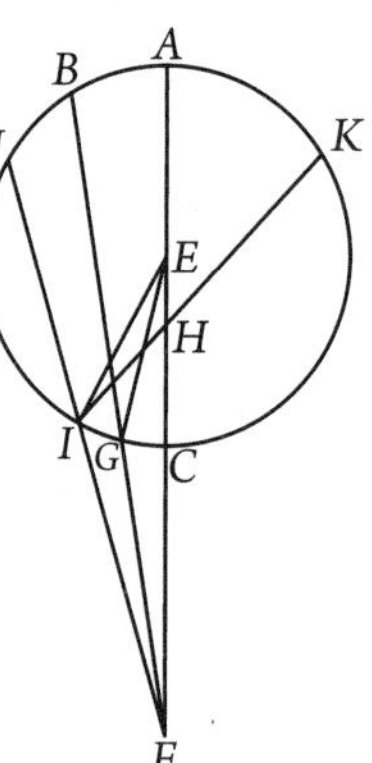

For, since in triangle *EIF* straight line *EC* is cut off not smaller than *EI*,
angle *CFI* will have a lesser ratio to angle *CEI* than straight line *EC* has to straight line *CF* {Lem. 12.1}.

And the ratio of *EC* to *CF* is not greater than the ratio of the speed of the epicycle to the speed of the star;
therefore angle *CFI* also will have a lesser ratio to angle *CEI* than the speed of the epicycle has to the speed of the star.

So that, since the star has been shown by us to be in forward motion wherever this happens,[1] we will find no arc of the epicycle and of the eccentric circle on which the star will appear in retrogradation.

H464

[The remainder of Book XII is omitted.]

[1] In Prop. 12.5 Ptolemy showed that if angle *GFI* : angle *IEG* :: epicycle's speed : star's speed, the star's compound motion will be forward (eastward). The same demonstration holds if *G* coincides with *C*.

PRELIMINARIES TO BOOK XIII

Context of the Following Selection

So far, in considering the hypotheses of the five planets, Ptolemy has treated them as if both the eccentric circles and the epicycles were in the plane of the mid-zodiac circle. In this final thirteenth book, Ptolemy takes up the problem of accounting for the planets' deviations from that circle, which in some cases can be quite considerable. He notes, first (Chapter 1), that the anomalies of latitude are twofold, zodiacal and heliacal, like the anomalies of longitude. After qualitatively describing the nature of these anomalies in the remainder of Chapter 1, he proceeds in Chapter 2 to set forth geometrical models to account for the motions.

The nature of these models is generally much the same for all planets. The eccentric circles are inclined to the mid-zodiac circle, each at its own angle and with its own region of maximum inclination. For Venus and Mercury, the eccentrics wobble in tune with the epicycles, while for Mars, Jupiter, and Saturn, the eccentrics are fixed in position. The epicycles are also inclined to the mid-zodiac circle, but rock from one side to the other about an axis that is perpendicular to the line drawn from the earth to the epicycle's center. This rocking is governed by a crank mechanism in the form of a small circle perpendicular to the plane of the mid-zodiac circle. This small circle is attached to the perigee of the epicycle, and moves the perigee northward and southward, perpendicular to the plane. Mercury and Venus have in addition a "slant" mechanism that rocks the epicycle around an axis directed from the earth to the epicycle's center. To make matters even more complicated, the centers of uniform motion of all these crank circles are different from the centers of the circles themselves: that is, they are all equipped with equants.

It is at this point that Ptolemy brings Chapter 2 to a close with the passage below. Evidently, he feared that readers would find the newly introduced mechanisms too fantastic to be accepted as physically possible. Thus, in reading statements such as, "it is not fitting even to judge what is simple (τὸ ἁπλοῦν) in itself in heavenly things on the basis of things that seem to be simple among us," readers must decide for themselves how far Ptolemy intended this critique to apply to the general planetary models, or whether he was specifically addressing the question of the plausibility of the mechanisms governing the latitudinal motions.

Book XIII

2. On the Mode of the Motion of the Hypothetical Inclinations and Obliquities.

H529

* * *

H532.12 And let no one believe that these kinds of hypotheses[1] are troublesome (ἐργώδης) when he considers the inadequacy of devices (ἐπιτεχνήματα) available to us. For it is not appropriate to compare what is human with what is divine, or to gain assurances concerning these great things from the most dissimilar examples. For what is more dissimilar than things "always the same" to things that are never the same, and things that would be hindered by anything[2] to those not even hindered by themselves?[3] But rather, it is fitting to attempt, to the greatest extent possible, to adapt simpler (ἁπλούστεραι) hypotheses to the motions in the heavens, and, should this not be viable, those that one can. For once all of the appearances are saved according to the H533 logical sequence of the hypotheses, would it still seem surprising to anyone that combinations of these kinds can happen to the motions of heavenly things? There is no hindering nature (φύσις κωλυτική) among them, but rather a nature commensurate, in terms of yielding and receding, with the natural motions of each of them, even if contrary motions occur; so that they all can pass through and appear through all their combinations (χύματα) without exception. And this occurs freely not only in particular circles, but also in the spheres themselves and the axes of revolutions. We see in the likenesses fashioned by us (αἱ κατασκευαζόμεναι παρ' ἡμῖν εἰκόνες[4]) that the

---

1 *these kinds of hypotheses*: the peculiarly complex mechanisms that Ptolemy used to account for the motions in latitude. See the Preliminaries to Book XII for an account of the mechanisms.

2 Being "hindered" by something is characteristic of the four sublunary elements, which are subject to "violent" motion and so may both offer and experience resistance and force.

3 Being "not even hindered by themselves" is characteristic of the fifth, supra-lunary element (*ether*), which is subject to no change and employs no force.

4 *the likenesses fashioned by us*: specifically, likenesses of the heavenly things. One example of such a likeness or model is the astrolabe. Ptolemy discusses this instrument in Book V, Chapter 1, "On Construction of an Astrolabe Instrument" (περὶ κατασκευῆς ἀστρολάβου ὀργάνου H350.13). On the topic of "images" (εἰκόνες), compare also Book I, Introduction {H6}.

combination and continuity of heavenly things in their different motions are troublesome (ἐργώδης) and hard to contrive (δυσπόριστος) as against unhindered motions; but in the heavens they are in no way ever hindered by this kind of mixture. Or rather, it is not fitting even to judge what is simple (τὸ ἁπλοῦν) in itself in heavenly things on the basis of things that seem to be simple among us, when for us the same thing is not at all similarly simple for everyone. For to those who consider the matter in this light, none of what happens in the heavens would seem simple, not even what is unchangeable in the primary motion; since the very fact that they are unchangeable for all time is, among us, not difficult (δύσκολον) but utterly impossible. But it is fitting to judge on the basis of the unchangeableness of the natures and motions in the heavens themselves. For in this way they would all appear even more simple than what seems simple among us, since no strain (πόνος) nor any annoyance (δυσχερεία) can be suspected concerning their periods. H53

[The rest of Book XIII is omitted]

APPENDIX 1
The Trigonometric Functions

Trigonometry is concerned for the most part with finding the magnitudes of certain sides and angles in triangles when the other sides and angles are given. It proceeds from general theorems, most of which Ptolemy uses, but which he does not usually formulate explicitly. Most of the computations in the *Almagest* that involve his "Table of Chords" (Book I, Chapter 11) are trigonometric. The table itself is nearly equivalent to a modern sine table.

1. Sine

Given any acute angle A, with CB dropped perpendicular from any point C on one of its sides to the other, we define the *sine* of A as the ratio of the side opposite A to the hypotenuse of the right triangle CBA. In the drawing, sin A (as we abbreviate it today) is $CB : CA$. In modern practice the sine of an angle is expressed as the quotient CB / CA.

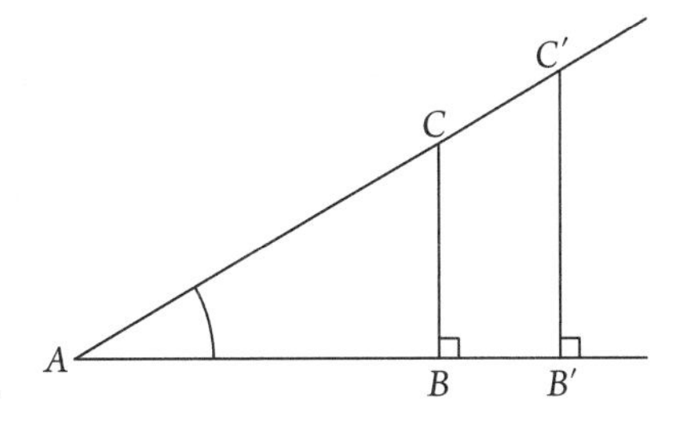

We can thus approximate the value of sin A by carrying out the indicated division to any number of places desired. The position of the perpendicular CB is arbitrary, since $CB / CA = C'B' / C'A$, where $C'B'$ is dropped perpendicularly from any other point C' on CA.

To see the relation of Ptolemy's Table of Chords to a sine table, take any arc PQ less than a semicircle, measured by the angle A at the center of the circle about O. Draw PO through to R. In Book I, Chapter 10, Ptolemy shows how to find a number n for the chord PQ such that $PQ / PR = n / 120$. If we then join QR, angle PQR will be right, since it is in a semicircle. Then by the definition of the sine,

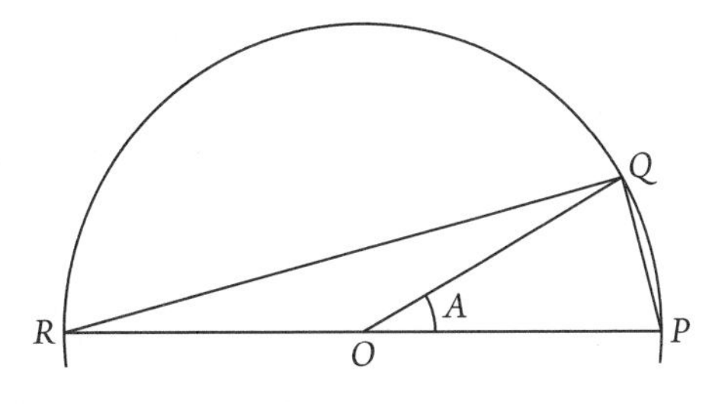

$$PQ / PR = \sin QRP = \sin \tfrac{1}{2}A \qquad \text{(Euclid III.20)}$$

Hence sin $\frac{1}{2}$A $= n / 120$. Thus, if we divide n (the length of the chord of a given arc) by 120, we will have the sine of an angle that is half the angle that measures that arc. Ptolemy's table from 0 to 180 degrees of arc thus becomes a sine table for angles from 0° to 90°.

The sine is called a *function* of the angle, because to every angle there corresponds a given sine; thus the value of the sine can be said to depend on the value of the angle. So too, Ptolemy's table was constructed by determining values of the chords for given arcs. Thus the chord is a function of the arc, as the sine is a function of the angle.

In using the table in the opposite sense, going from the chord to the arc, we are also speaking of a functional relationship: instead of asking, What is the chord of the arc of A degrees? we are asking, What is the arc whose chord is n parts? As the chord was a function of the arc, so the arc is a function of the chord. Since the same values are paired, but in reverse order, we speak of the one function as the inverse of the other: they are two aspects of the same relationship.

2. Cosine

The cosine of the angle A in the drawing is defined as the ratio between the side adjacent to A and the hypotenuse; that is, $AB:AC$. Its numerical expression is the quotient AB/AC. Ptolemy does not use the cosine but derives its equivalent in each case. Now AB/AC is also the sine of the angle C, so that the relation between the sine and the cosine, its co-function, will be:

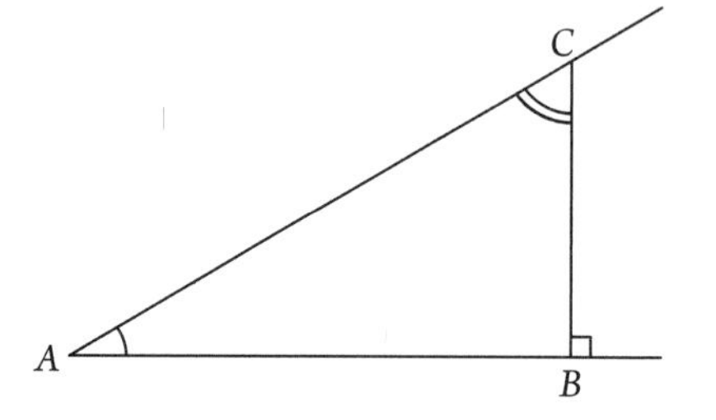

$$\cos A = \sin(90° - A),$$
$$\sin A = \cos(90° - A).$$

Thus to derive a table of cosines from the sine table we have only to read the sine table backwards! For, by the above relations, $\cos 1° = \sin(90° - 1°) = \sin 89°$; $\cos 2° = \sin(90° - 2°) = \sin 88°$; and so on.

Two angles that sum to 90° are said to be *complements* of one another. From the above, we see that the cosine of A equals the sine of the complement of A. Indeed, that is the origin of the name "cosine": it is derived from the Latin *complimenti sinus*, the "sine of the complement." A similar observation holds for other co-functions.

3. Tangent and cotangent

These are another important and convenient pair of trigonometric functions. Ptolemy does not refer to either of them explicitly. In the drawing above, the *tangent* of the angle A (abbreviated $\tan A$) is defined as the ratio of the side opposite A to the side adjacent to A, that is, $CB : BA$. Its numerical expression is CB / BA, the quotient of the lengths of the two sides. Similarly, the *cotangent* of A (abbreviated $\cot A$) must be the tangent of C, the complement of A; therefore $\cot A$ is defined as the ratio $AB : BC$. Its numerical expression is the quotient AB / BC.

Notice in the same drawing that CB/BA, the tangent of A, corresponds to the quotient of the gnomon's height by the length of its shadow. Measurements made with the gnomon therefore give us the tangent of the sun's altitude, from which the altitude itself may be found by means of a table of tangents.

A table of tangents, in turn, can easily be derived from the tables of sines and cosines; for it is easy to show that CB/BA is just the quotient of $\sin A$ by $\cos A$. Similarly, the cotangent of A is seen to be the quotient of $\cos A$ by $\sin A$.

A table of cotangents can also be obtained by reading the table of tangents backwards, since the cotangent is the co-function of the tangent. This figure shows the four functions in relation to the unit circle. Because the radius is taken as 1, $\sin A$, $\cos A$, $\tan A$, and $\cot A$ can all be represented as single lines. (To see why, look for similar triangles.) The figure makes it clear that the sine of A equals half the chord of $2A$; it also shows how the tangent function got its name.

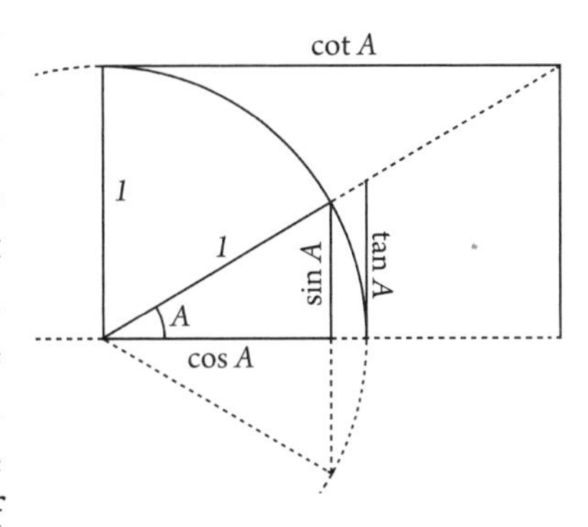

4. Origin of the trigonometric functions

The rationale for the trigonometric functions becomes clear in the course of working through Ptolemy's calculations; for in carrying them out one often has first to double an arc, then find the chord of that doubled arc, and finally take half the chord. Clearly it would be handier to have a table of half-chords of double-arcs; and within a few centuries of Ptolemy's work, Indian mathematicians introduced such a table. At the same time, it was recognized that it would be useful to have a separate table for the line that goes out from the angle to the midpoint of the double arc—which is simply the sine of the complement of the angle, the sine of $(90° - A)$. These became our sines and cosines, respectively.

The other functions are later inventions. The cotangent and tangent were originally introduced by Islamic mathematicians in tables giving the lengths of shadows at different elevations of the sun. Their modern names were introduced by Copernicus's student G. J. Rheticus in a book published in 1596; however, the construction of the tangent as in the figure above was first proposed by Abu'l-Wafā al-Buzjani in the tenth century. Indeed, the figure as a whole is essentially the same as Abu'l-Wafā's diagram.

APPENDIX 2
Days of the Year

1. Ordinary Years

	Jan	Feb	Mar	Apr	May	Jun	Jul	Aug	Sep	Oct	Nov	Dec
1	1	32	60	91	121	152	182	213	244	274	305	335
2	2	33	61	92	122	153	183	214	245	275	306	336
3	3	34	62	93	123	153	184	215	246	276	307	337
4	4	35	63	94	124	155	185	216	247	277	308	338
5	5	36	64	95	125	156	186	217	248	278	309	339
6	6	37	65	96	126	157	187	218	249	279	310	340
7	7	38	66	97	127	158	188	219	250	280	311	341
8	8	39	67	98	128	159	189	220	251	281	312	342
9	9	40	68	99	129	160	190	221	252	282	313	343
10	10	41	69	100	130	161	191	222	253	283	314	344
11	11	42	70	101	131	162	192	223	254	284	315	345
12	12	43	71	102	132	163	193	224	255	285	316	346
13	13	44	72	103	133	164	194	225	256	286	317	347
14	14	45	73	104	134	165	195	226	257	287	318	348
15	15	46	74	105	135	166	196	227	258	288	319	349
16	16	47	75	106	136	167	197	228	259	289	320	350
17	17	48	76	107	137	168	198	229	260	290	321	351
18	18	49	77	108	138	169	199	230	261	291	322	352
19	19	50	78	109	139	170	200	231	262	292	323	353
20	20	51	79	110	140	171	201	232	263	293	324	354
21	21	52	80	111	141	172	202	233	264	294	325	355
22	22	53	81	112	142	173	203	234	265	295	326	356
23	23	54	82	113	143	174	204	235	266	296	327	357
24	24	55	83	114	144	175	205	236	267	297	328	358
25	25	56	84	115	145	176	206	237	268	298	329	259
26	26	57	85	116	146	177	207	238	269	299	330	360
27	27	58	86	117	147	178	208	239	270	300	331	361
28	28	59	87	118	148	179	209	240	271	301	332	362
29	29		88	119	149	180	210	241	272	302	333	363
30	30		89	120	150	181	211	242	273	303	334	364
31	31		90		151		212	243		304		365

2. Leap Years

	Jan	Feb	Mar	Apr	May	Jun	Jul	Aug	Sep	Oct	Nov	Dec
1	1	32	61	92	122	153	183	214	245	275	306	336
2	2	33	62	93	123	153	184	215	246	276	307	337
3	3	34	63	94	124	155	185	216	247	277	308	338
4	4	35	64	95	125	156	186	217	248	278	309	339
5	5	36	65	96	126	157	187	218	249	279	310	340
6	6	37	66	97	127	158	188	219	250	280	311	341
7	7	38	67	98	128	159	189	220	251	281	312	342
8	8	39	68	99	129	160	190	221	252	282	313	343
9	9	40	69	100	130	161	191	222	253	283	314	344
10	10	41	70	101	131	162	192	223	254	284	315	345
11	11	42	71	102	132	163	193	224	255	285	316	346
12	12	43	72	103	133	164	194	225	256	286	317	347
13	13	44	73	104	134	165	195	226	257	287	318	348
14	14	45	74	105	135	166	196	227	258	288	319	349
15	15	46	75	106	136	167	197	228	259	289	320	350
16	16	47	76	107	137	168	198	229	260	290	321	351
17	17	48	77	108	138	169	199	230	261	291	322	352
18	18	49	78	109	139	170	200	231	262	292	323	353
19	19	50	79	110	140	171	201	232	263	293	324	354
20	20	51	80	111	141	172	202	233	264	294	325	355
21	21	52	81	112	142	173	203	234	265	295	326	356
22	22	53	82	113	143	174	204	235	266	296	327	357
23	23	54	83	114	144	175	205	236	267	297	328	358
24	24	55	84	115	145	176	206	237	268	298	329	259
25	25	56	85	116	146	177	207	238	269	299	330	360
26	26	57	86	117	147	178	208	239	270	300	331	361
27	27	58	87	118	148	179	209	240	271	301	332	362
28	28	59	88	119	149	180	210	241	272	302	333	363
29	29	60	89	120	150	181	211	242	273	303	334	364
30	30		90	121	151	182	212	243	274	304	335	365
31	31		91		152		213	244		305		366

APPENDIX 3
The Menelaus Theorems

Statement of the Menelaus Theorems is greatly facilitated by adopting the locution of *internal* and *external* division of a line. In the following figure, point K lies on line AB and divides it into segments AK and BK. This is the ordinary case of division of a line by a point; let us call it *internal* division.

But consider point E, which lies outside line AB but collinear with it. Let us now say that point E divides line AB *externally*, producing segments AE and BE. All cases of the planar Menelaus Theorems may then be conveniently stated as follows: *If a straight line cuts the sides of a triangle either internally or externally, the ratio of segments of any side is the compound of the ratios of segments of the other two sides, respectively, taken in the same order.*

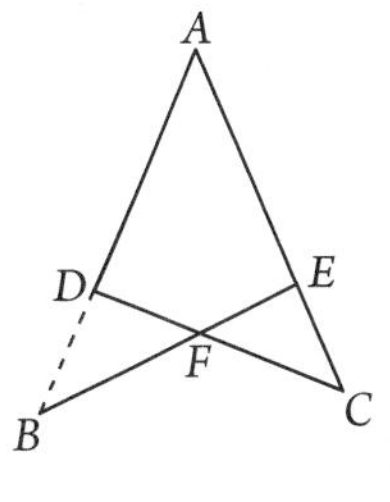

The order in which segments are taken is determined by the order of the first proportion. In the adjacent diagram, in triangle ACD, the order of segments of side CA may be taken as CE, EA or AE, EC. The order of segments on the remaining sides must follow the order on CA, proceeding either from C to D to A, or from A to D to C. .

For example, to find the ratios whose compound is $CE{:}EA$, first take segments CF, FD on CD and then segments DB, BA on DA. Then a Menelaus theorem states that

$$CE : EA :: CF : FD \text{ comp. } DB : BA,$$

corresponding to Ptolemy's Lemma 1.4 in Book I, Chapter 13.

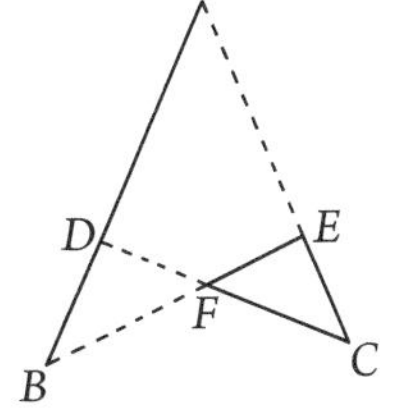

In the second diagram, points of division of the sides of triangle CEF are all external. Proceeding from C to E, in a similar way we have a Menelaus theorem stating that

$$CA : AE :: CD : DF \text{ comp. } FB : BE,$$

which corresponds to Ptolemy's Lemma 1.3.

A corresponding locution of internal and external division of *arcs of a great circle* will similarly facilitate expression of the spherical Menelaus Theorems.

Select Bibliography

The standard edition of the *Almagest* is:

Ptolemy, Claudius, *Syntaxis mathematica,* 2 vols., Ed. J. L. Heiberg (Teubner, 1898–1903). Available online through www.wilbourhall.org.

There are two complete English translations:

Ptolemy, Claudius, *The Almagest by Ptolemy,* trans. by R. C. Taliaferro, in *Great Books of the Western World,* 16 (Encyclopædia Britannica, 1952).

This translation was originally commissioned and published by St. John's College, Annapolis, in 1937.

Ptolemy's Almagest, trans. by G. J. Toomer (Springer Verlag, 1984).

Two valuable works on Ptolemy's astronomy are:

Pedersen, Olaf, *A Survey of the Almagest* (Odense University Press, 1974)

Taub, Liba, *Ptolemy's Universe: The Natural Philosophical and Ethical Foundations of Ptolemy's Astronomy* (Open Court, 1993)

The most accessible and informative general work on ancient astronomy is

Evans, James, *The History and Practice of Ancient Astronomy* (Oxford University Press, 1998).

Rather difficult to read, but the ultimate authority, is:

Neugebauer, Otto, *History of Ancient Mathematical Astronomy,* 3 vols. (Springer Verlag, 1975).

Two works of Euclid are cited in this volume:

Heath, T. L., *The Thirteen Books of Euclid's Elements,* 3 vols., (Cambridge University Press, 1908; reprinted Dover Publications, 1956)

Euclid, *Euclid's Elements: all thirteen books complete in one volume* (Green Lion Press, 2013)

Euclid, *The Data of Euclid,* trans. by G. L. McDowell and M. A. Sokolik (Union Square Press, 1993)

For the writings of Parmenides (cited in p. 25n1), see:

A Presocratics reader: selected fragments and testimonia, edited, with introduction, by Patricia Curd; translations by Richard D. McKirahan. (Hackett, 2011).

Index

About the Author, Claudius Ptolemy

Most of what is known about Ptolemy's life is derived from his works. From his observations reported in the *Almagest* and other works, it appears that he was born about 100 C.E. and lived until at least 170. All his work was done in Alexandria. His second most famous achievement was his *Geography* in which he mapped the entire inhabited world, devised methods for mapping, used the equivalent of latitude and longitude lines, and developed the projection of the spherical globe onto two dimensional maps. In addition to the *Almagest,* Ptolemy wrote *Hypotheses of the Planets* and other shorter astronomical works. In true Alexandrian fashion, he also composed epochal books on astrology and optics.

About the Translator, Bruce M. Perry

Bruce Perry is a classicist and Sanskrit scholar who has taught in the integrated Great Books program at St. John's College, Santa Fe, since 1990. After receiving his Ph.D. in classics from the University of Washington in 1983, he went on to study and teach at the University of Pennsylvania before joining the St. John's faculty. At St. John's he has taught classes on the *Almagest* and his translation has become the version preferred by students and faculty.

About the Editor, William H. Donahue

William Donahue is a historian of astronomy known chiefly for his studies and translations of the work of Johannes Kepler. He received his Ph.D. in 1973 from the University of Cambridge, for a dissertation on *The Dissolution of the Celestial Spheres, 1595–1650,* which was published in 1981 by Arno Press. From 2005 to 2016 he was on the faculty of St. John's College, Santa Fe, where he was Director of Laboratories. He and his wife Dana Densmore are owners and Co-Directors of Green Lion Press.